Table of Contents

Executive Summary 1

Introduction 3

Background 5
   Role of Hydrogen 6

Challenges 11
   Technology Challenges 12
   Public and Private Sector Organizational Challenges 19
   Commercial Sector 20
   Safety and Emergency Response Challenges 22
   Sustained Commitment Challenges 24

Federal Progress to Date 31
   DOT Research and Achievements 31
   DOE Research and Achievements 33
   USDA Research and Achievements 36
   EPA Research & Achievements 36
   Federal Interagency Coordination 37

Constructing an Infrastructure Plan 39
   Recent Studies Related to a Hydrogen Infrastructure 39
   Where is a National Plan? 41

The Future 43

Appendix A – Infrastructure Maps 45

Appendix B – Comparison of Alternative Fuels Technical Readiness 49

Appendix C – International Hydrogen and Fuel Cell Infrastructure
                Development for Transportation Applications 51

Appendix D – List of Nation's Hydrogen Stations 55

Acknowledgements

This report is a high level overview of the infrastructure challenges of a transition to a hydrogen fueled transportation system and the Federal efforts needed to mitigate and eliminate them. It is meant to provide a synopsis of the substantial work that has been done over the last several years by the Department of Transportation, Department of Energy, DOE's National Laboratories, National Academy of Sciences' National Research Council, the Department of Agriculture, the National Hydrogen Association, the California Fuel Cell Partnership and other key stakeholders.

The Department of Transportation is pleased to be at the forefront of the effort and thanks its Federal and other colleagues for sharing their research to make this report possible.

Executive Summary

In book two of the conference report on Public Law 111-8, the Omnibus Appropriations Act, 2009, Congress directed the Department of Transportation's (DOT's) Research and Innovative Technology Administration (RITA) "to provide a report to the House and Senate Committees on Appropriations within 90 days of the enactment of this Act detailing the challenges of installing hydrogen infrastructure. This report should include a comprehensive plan to increase the number of hydrogen fueling stations around the country, focusing on the regions with greatest demand and need. The agency is instructed to coordinate with the Department of Energy to complete this report."

Building a national infrastructure for hydrogen will not be an easy task. It will take a dedicated long-term focus by government and the private sector. As one participant at the National Academies of Science, National Research Council (NRC)'s Summit on America's Energy Future indicated, the country is at the beginning of a 30-year planning window (from 2010 through 2040) for putting in place the policies, technologies and infrastructures needed to meet the nation's mobility needs through the 22nd century.[1]

In the past few years, DOT, the Department of Energy (DOE), and their industrial and academic partners have made significant advancements in putting hydrogen technologies on the path to validation and eventual commercialization. Notable improvements were reported and independently verified for the performance and costs of fuel cells, the capacity of on-board hydrogen storage, and hydrogen fueling technology.

There is no single national plan for building a national infrastructure. Instead, there are a variety of national plans orchestrating each of the numerous activities that constitute a hydrogen infrastructure. This is due, in large part, to the dispersal of authority and responsibility for all of these elements across the public and private sectors. Coordination efforts are focused on the process owners and stakeholders who make the largest difference in achieving disparate goals. However, there are national, and sometimes international, efforts addressing each of the key aspects such as safety codes and standards, Federal research and development, requirements for station siting, and providing outreach and tools for State and local decisionmakers. The work that is being done today is providing the context that will make a national framework possible.

Supported by this work, senior decisionmakers face choices as they reconcile and integrate these and other accomplishments into a path forward for an alternatively fueled America. One important task will be to harmonize all of the short, medium and long-term solutions this transition involves. However, this is not solely a Federal responsibility. State, local and private sector stakeholders will be key to leveraging and realizing a common commitment for this fundamental change in American mobility. Accomplishing this transition will be no less impressive than building a transcontinental railroad or the Interstate highway system.

Key challenge areas decisionmakers face include:

▶ **Technology**

- *innovations* to increase the supply, efficiency, range and cost competitiveness of fuel cell vehicles, and reduce the cost of producing hydrogen from domestic resources using green production methods.

▶ **Public & private sector organizational**

- *Land use and station siting* guidance to ensure the safe and efficient development of this new infrastructure including development of future improvements to reduce the size of the current station footprint.

- *Public education and outreach* to increase awareness, motivate key stakeholders, and facilitate the acceptance of the new technology.

---

[1] National Academies of Science, National Research Council, Summit on America's Energy Future: Summary of a Meeting and Transitions to Alternative Transportation Technologies – A Focus on Hydrogen

- **Commercial sector**
  - *Market development and deployment* including policy decisions about whether implementation should focus on growing urban and regional markets where there is likely to be strong consumer demand or on a national network so that vehicles can operate regardless of location.
  - *Partnerships* to bring together the stakeholders whose collaboration is essential to the deployment of hydrogen vehicles and a hydrogen infrastructure, i.e., Federal, State, and local government, automakers, fuel providers, electricity producers, other relevant industries, academia, environmental groups, and the public.

- **Safety codes & standards**
  - **Universally accepted requirements** to establish the appropriate safety, quality and consumer protection also be provided to match fossil fuel standards including the safety of compressed hydrogen (CH2) and liquid hydrogen (LH2) fueled vehicles and subsystems, of fueling infrastructure and of fueling interfaces, as well as safe integration and compatibility with mixed fleet and fuels operations during a long transition period.
  - ***Emergency response training*** to provide the knowledge and tools first responders will need to deal with the different dangers hydrogen presents as well as provide the regulatory requirements needed to address the new technologies and innovations this transition will generate.

- **Sustained commitment**
  - ***Programs and incentives*** to address the expected cost differentials between hydrogen vehicles and conventional vehicles during the transition period. Some of these activities should be coordinated with the safety, codes and standards activities in order to accelerate the insurance industry's adoption of comparable rate structures and procedures.

As for identifying a network of hydrogen stations, there is much future work to be done. From a systemic perspective, NRC envisions that by 2050 there could be 220 million hydrogen fuel cell vehicles, 1,200 to 1,800 hydrogen refueling stations, 210 central plants, and 80,000 miles of pipeline.[2] Today, DOE estimates there are about 60 hydrogen refueling stations across the nation. The most active effort to create this infrastructure is the California Fuel Cell Partnership's program to create 41 stations within its state by 2015.[3]

---

[2] National Academies of Science, National Research Council, Summit on America's Energy Future: Summary of a Meeting and Transitions to Alternative Transportation Technologies – A Focus on Hydrogen
[3] *California Fuel Cell Action Plan.*

Introduction

In book two of the conference report for Public Law 111-8, the Omnibus Appropriations Act, *2009*, Congress directed the Department of Transportation's Research and Innovative Technology Administration (RITA) "to provide a report to the House and Senate Committees on Appropriations within 90 days of the enactment of this Act [March 11 – June 11] detailing the challenges of installing hydrogen infrastructure. This report should include a comprehensive plan to increase the number of hydrogen fueling stations around the country, focusing on the regions with greatest demand and need. The agency is instructed to coordinate with the Department of Energy to complete this report."

This report is a synthesis of already published work, as well as an identification of the policy decisions and commitments needed to make hydrogen fuels an integral part of how this nation moves its citizens and commercial goods. It is organized to: 1) set the context of why hydrogen is important to transportation's transition from fossil fuels and to the nation's climate change mitigation efforts, 2) explore the major challenge areas this transition presents, 3) discuss the Federal progress made to date, 4) highlight the studies that underpin today's strategic thinking on the topic, and 5) identify key areas for future action.

Background

Building the nation's rail and highway infrastructures were not easy tasks. It took decades of commitment and investment by government and the private sector. As summarized in figure 1.1, the nation's railroad system took over three decades of active Federal investment through land grants. The national road system took three decades of Federal planning and three decades of Federal investment. The first automobile was manufactured in the United States in 1893. Through strictly private investment, it took more than 25 years to establish a national network of fueling stations.

Figure 1.1: Building the U.S. Rail & Highway Infrastructure

- **1830** First track laid in US
- **1832** First train runs in US
- **1850** Congress Begins Land Grants
- **1850 – 1871** Congress Grants 143M acres to Rails For transcontinental system - 12% of Total Lands granted
- **1869** First transcontinental line completed
- **1871** Congress End Land Grants
- **1893** First car made in US
- **1907/8** First gas stations open in US
- **1924 - 1926** First National Road System Planned for US
- **1924 - 1991** National Road System Planned and Interstate System completed Total Cost $128.9B, Federal Share $114.3B
- **1956** Congress creates Interstate system
- **1957 - 1991** Instate System constructed

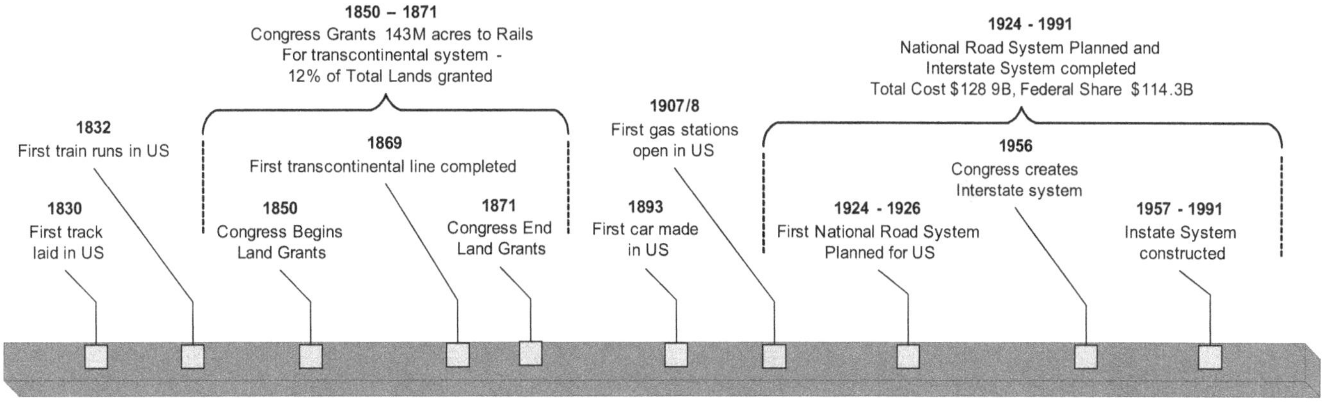

Source: U.S. Department of Transportation, Research and Innovative Technology Administration, 2009.

ROLE OF HYDROGEN

The national interest in creating a hydrogen infrastructure reflects the potential role hydrogen could play in meeting the goal of reducing greenhouse gas (GHG) emissions by mid-century. Without hydrogen and hydrogen fuel cells helping to power transportation, the Department of Energy (DOE) estimates that attaining this goal will slip from mid-century to about 2075 or later.

As shown in figure 1.2, the Congressional Research Service (CRS) estimates 63 percent of GHG come from fuel combustion and 98 percent of transportation fuel is fossil-based. For 2008, the Energy Information Administration estimates the total for fossil-based transportation fuel at 99 percent.

Figure 1.2: Greenhouse Gas Emissions in 2005

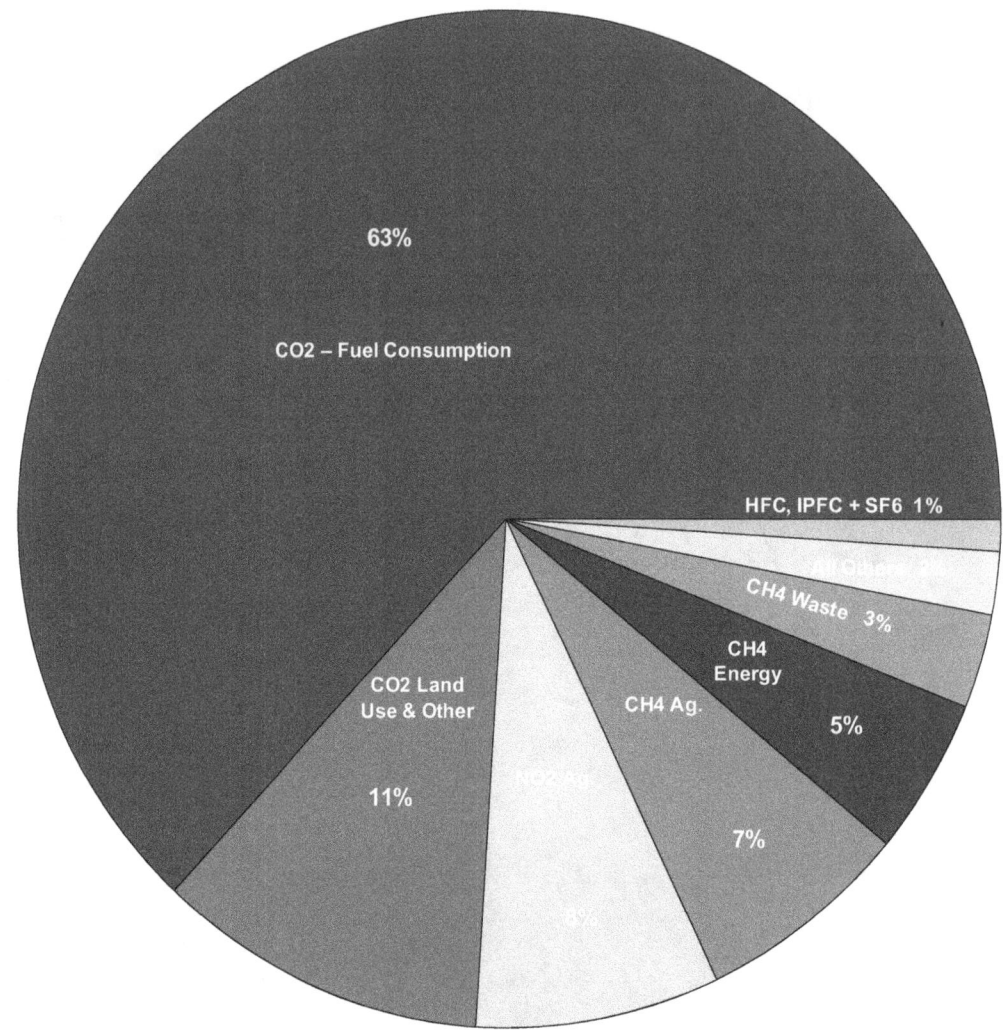

Source: Congressional Research Service graphic with estimates from International Energy Agency CO2 Emission from Fuel Combustion 1971-2005, 2007, on-line database as of Jan. 16, 2008.

6 Challenges of Building a Hydrogen Infrastructure

In figure 1.3 below, the State of California relies on super ultra low carbon vehicles and hydrogen to improve its air quality.

Figure 1.3: Strategies to Reduce Greenhouse Gases

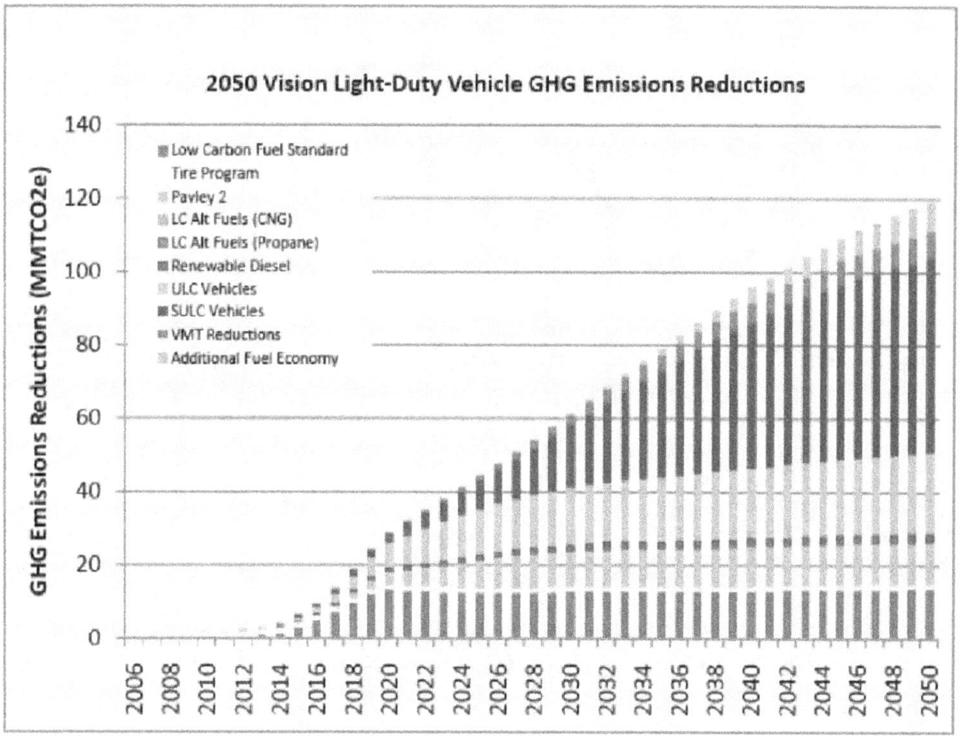

Source: California Energy Commission, November 2008.

A clearer view of hydrogen's potential contribution is shown in figures 1.4 and 1.5. In figure 1.4, the National Research Council (NRC) chart shows the role hydrogen fuel cell vehicles potentially could play between 2025 and 2050 in reducing GHG emissions. In this optimistic scenario, NRC sees hydrogen fueled vehicles having greater VMT than hybrids after 2040 and being the predominant vehicle fuel by 2060. In figure 1.5, DOE shows the results of its best case scenario modeling for the adoption of hydrogen fuel cell vehicles. NRC and DOE estimate about 200 million to 220 million hydrogen vehicles in operation by 2050.

Figure 1.4: Hybrids' and Hydrogen's Share of Projected Vehicle Miles Traveled (VMT)

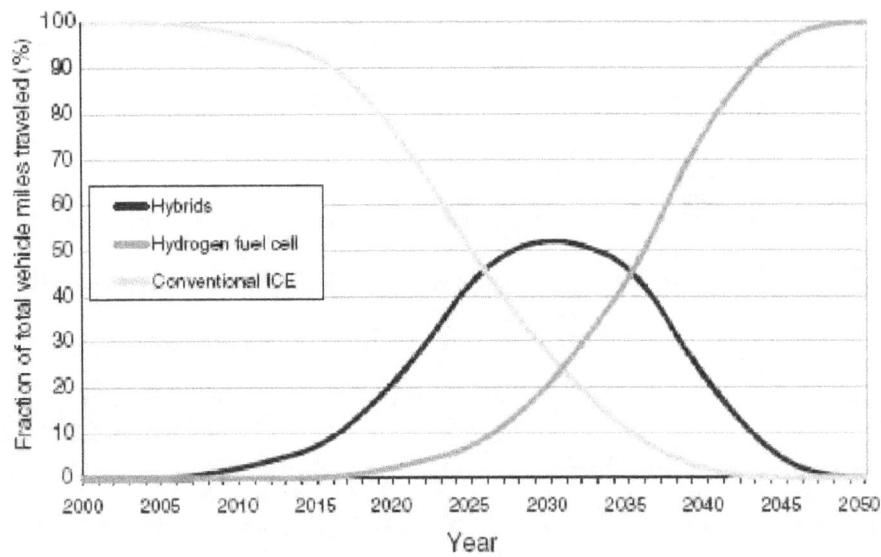

Source: National Academy of Sciences, 2008. *The National Academies Summit on America's Energy Future: Summary of a Meeting*, p. 75. The National Academies Press, Washington, DC.

Figure 1.5: DOE Scenarios Showing the Adoption of Hydrogen Fuel Cell Vehicles

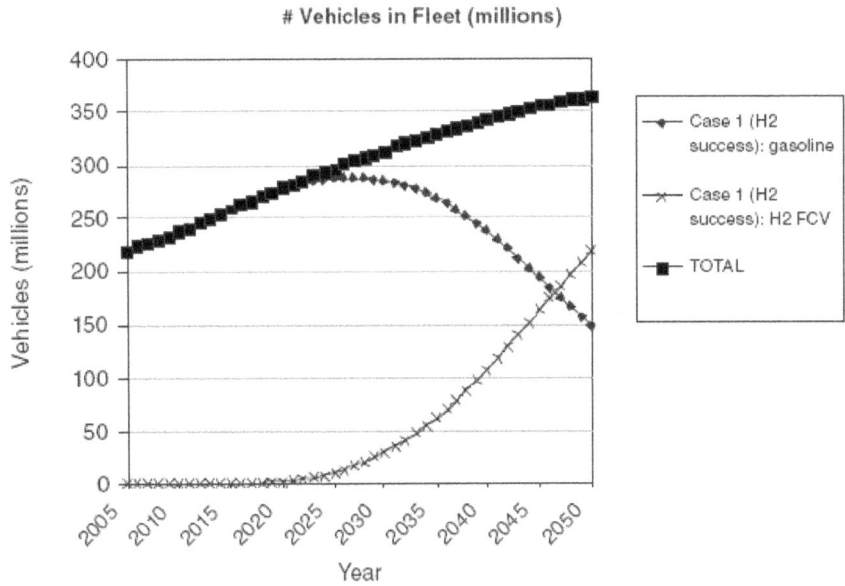

Source: National Research Council. 2008. *Transitions to Alternative Transportation Technologies—A Focus on Hydrogen*, p. 8. The National Academies Press, Washington, DC.

During its *Summit on America's Energy Future* in Fall 2008, NRC described the nation as being at a critical juncture for planning how we will meet the mobility needs of this and future generations. As one conference participant illustrated in figure 1.6, the country is at the beginning of a 30-year planning window (from 2010 through 2040) for putting in place the policies, technologies and infrastructures needed to meet the nation's mobility needs through the 22nd century.

Figure 1.6: Timeframe to Achieve Long-Term Emissions Reduction Outcomes

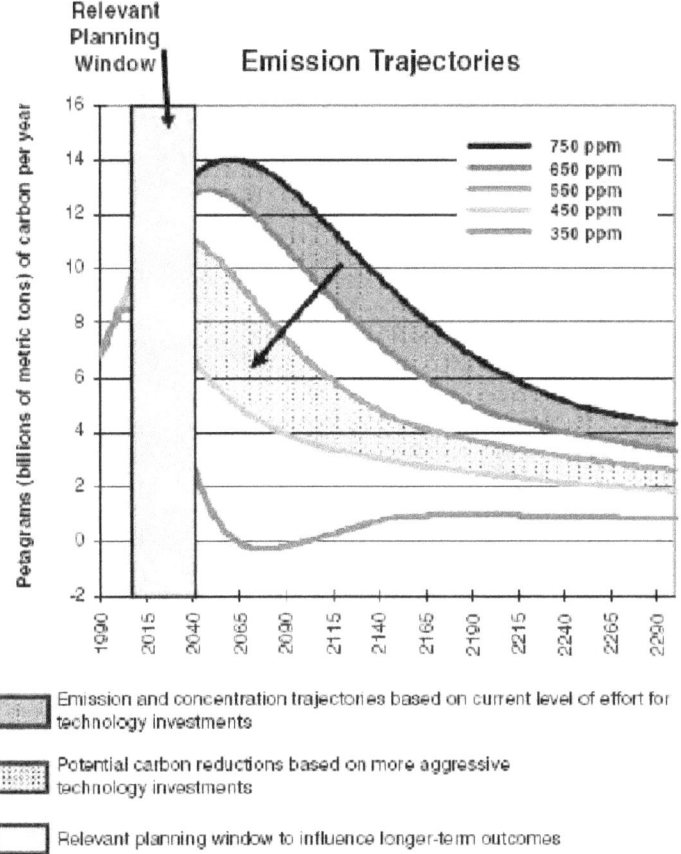

Source: National Academy of Sciences. 2008. *The National Academies Summit on America's Energy Future: Summary of a Meeting*. p. 112. The National Academies Press, Washington, DC.

Challenges

Growing the current national retail network of about 60 hydrogen fueling stations to achieve a viable hydrogen infrastructure will take significant investment and technological innovation. Figure 2.1 shows a 2008 snapshot of hydrogen fueling stations. The figure of stations nationwide is approximated at 60 because stations are opening and closing in response to market forces. Appendix A outlines some of the transmission and distribution networks that could help support a hydrogen infrastructure. Table 2.4 at the end of this section summarizes the challenges a hydrogen transition entails. The table in Appendix B shows the challenges facing all non-fossil fuels.

Figure 2.1: Listing of Hydrogen Stations Nationwide

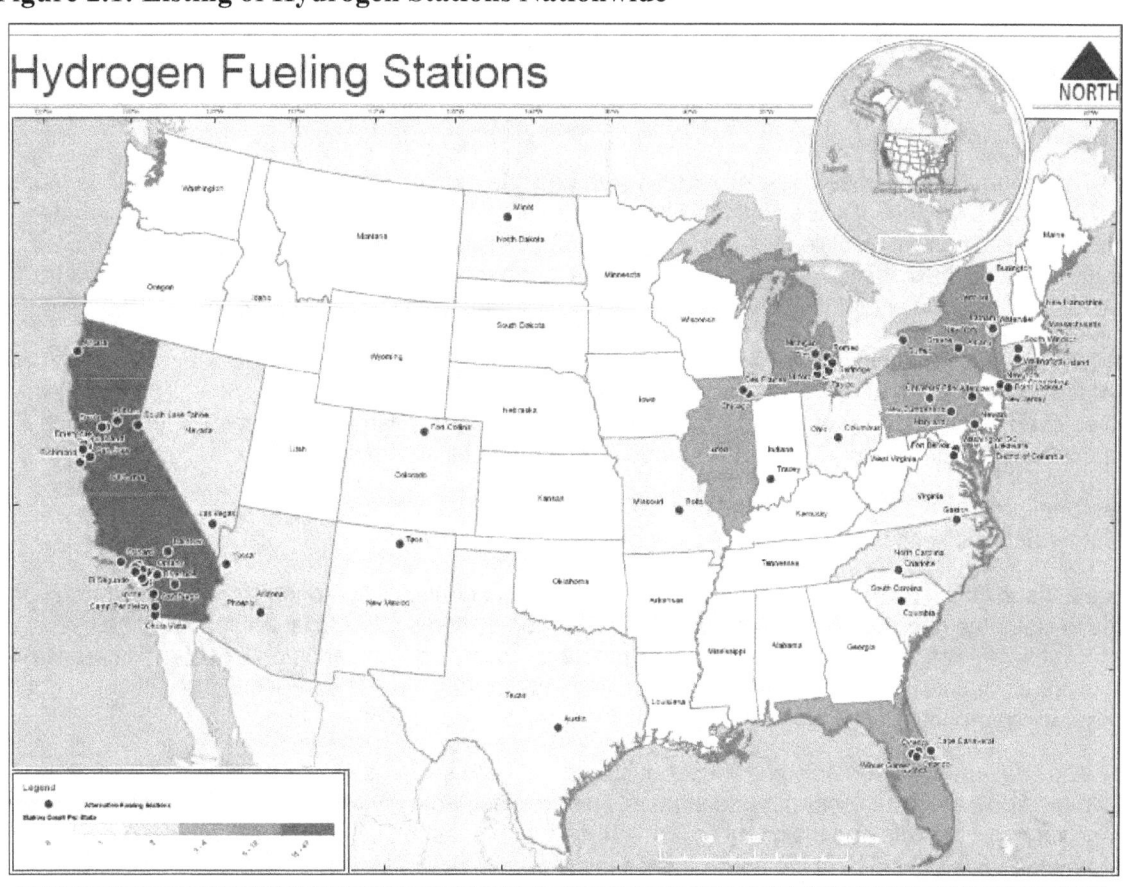

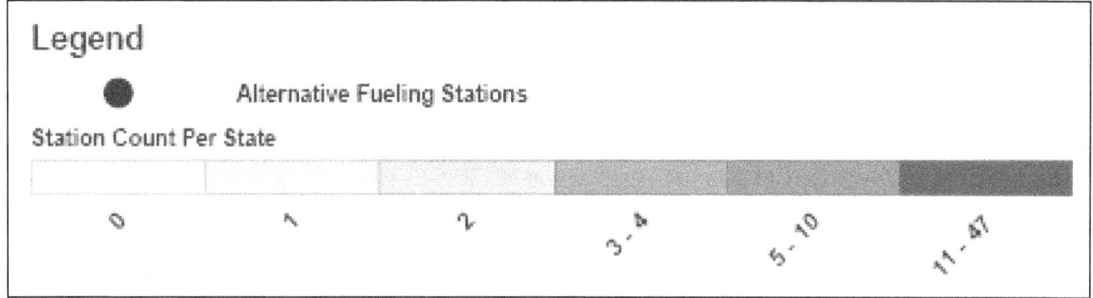

Source: U.S. Department of Energy, National Renewable Energy Laboratory, available at http://www.afdc.energy.gov/afdc/fuels/hydrogen_locations.html?print as of Feb. 16, 2010.

Technology is just one of the challenges a transition to hydrogen faces. There also are public and private sector constraints, commercial issues, safety codes and standards that need to be implemented, safety and emergency response trainings to be developed, as well as sustained commitment to the goal. This section explores each of these areas, the progress that has occurred to date, as well as the remaining barriers to be overcome and/or strategies that need to be decided.

TECHNOLOGY CHALLENGES

If the country is to reduce significantly the greenhouse gases that threaten the environment, technological innovations in a variety of areas will be necessary. Current Federal research, development and deployment programs are playing key roles in this transition.

A hydrogen infrastructure faces three important technology challenges:

- Developing a fuel cell that can match a traditional car's useful life and performance.

- Improving storage capacity, performance and rate of refueling for a vehicle's on-board fuel supply.

- Closing the gap between the cost of hydrogen vehicles and fuel when compared to traditional cars and gasoline.

Fuel Cell Performance

The NRC reported that fuel cell "stack" operating life has grown from about 1,000 hours in 2004 to more than 1,500 hours. In 2008, DOE's National Renewable Energy Laboratory documented an independent validation of 140 fuel cell vehicles that showed nearly 2,500 hours under real world conditions.[1] More recently, fuel cells have lasted as long as 7,300 hours in laboratory testing.[2] The Federal goal is 5,000 hours, which is equivalent to 150,000 miles of engine life for gasoline-powered vehicles.

Fuel cells also are beginning to match the distances between refuelings for gasoline vehicles – about 300 miles. NRC noted that a 2007 Toyota test did exceed the 300-range goal. However, the NRC reported, it did so at an estimated cost of $15 to $18 per kilowatt hour for on-board fuel storage systems, which is much higher than the goal of $2 per kilowatt hour for commercial success.

These benchmarks are important because, according to the NRC, they are the performance measures for consumer and market acceptance of this technology.

Figure 2.2 shows the gaps between current performance and goals for hydrogen storage. Research continues to close the gaps between the different compression systems and pressures for refueling hydrogen fuel cell vehicles or HFCVs. The amount of progress toward goal depends on whether compression agents are cryogenic, chemical or some other compound. To date, liquid hydrogen is the closest to goal, followed by cryo-compressed fuel cells using the higher pressure rate of 700 bars.

NRC also cautioned that the size and weight of current fuel cell systems must be further reduced to match fossil-fueled operating efficiency. It added that this also applies not only to the fuel cell stack, but also to the ancillary components and major subsystems (e.g., fuel processor, compressor/expander, and sensors) making up the balance of the power system.

In its report, *Transitions to Alternate Transportation Technologies – a Focus on Hydrogen,* the NRC assessed the current state of fuel cell technology as:

> "Lower-cost, durable fuel cell systems for light-duty vehicles are likely to be increasingly available over the next 5-10 years and, if supported by strong government policies, commercialization and growth of HFCVs [hydrogen fuel cell vehicles] could get underway by 2015, even though all DOE targets for HFCVs may not be fully realized."

---

[1] DOE 2008 Annual Report http://www.hydrogen.energy.gov/pdfs/progress08/v_c_1_debe.pdf ; National Renewable Energy Laboratory's latest durability Controlled Demonstration Project (CDP): http://www.nrel.gov/hydrogen/docs/cdp/cdp_1.ppt and completed 2008 CDPs: http://www.nrel.gov/hydrogen/pdfs/44256.pdf slide 4, also the "max projection."
[2] http://www.hydrogen.energy.gov/pdfs/progress08/v_c_1_debe.pdf

Figure 2.2: Hydrogen Storage Gaps: Status vs. Targets

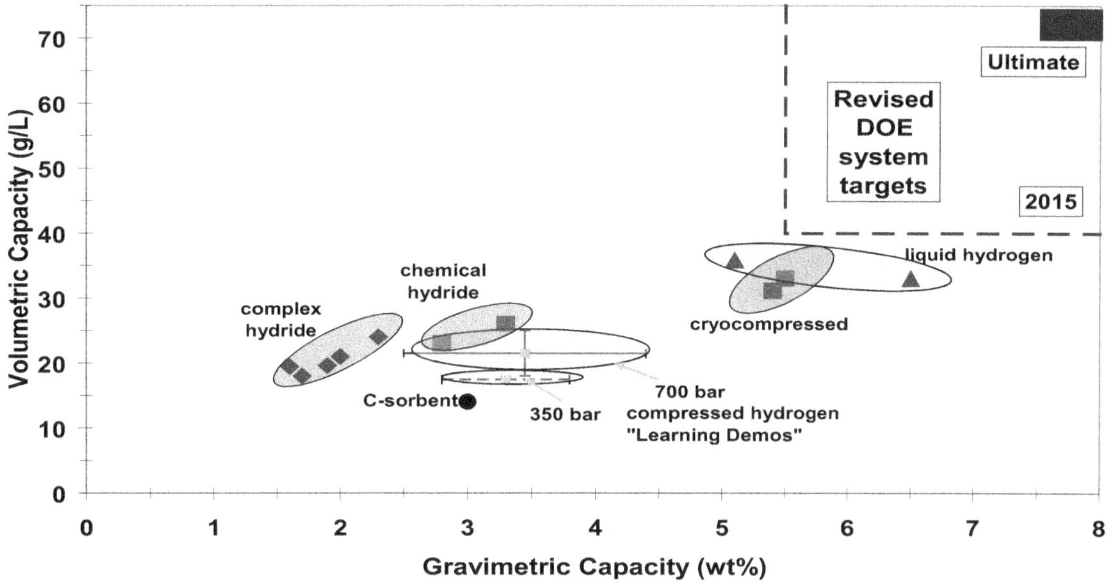

Source: U.S. Department of Energy. 2008. Office of Efficiency and Renewable Energy's Fuel Cell Technologies Program, *FY 2008 Annual Progress Report*, Hydrogen Program, p. 439, available at http://www.hydrogen.energy.gov/annual_progress08.html as of Feb. 16, 2010.

Fuel and Vehicle Supply

Today, alternative fuels account for only 2 percent of the nation's transportation fuels. According to research done by RITA's Volpe National Transportation Systems Center, alternative fuels in 2006 provided about 5 billion of the roughly 184 billion gasoline equivalent gallons needed to move our citizens and commerce. Most transportation fuels are destined for the nation's 238.1 million light vehicles and 9.3 million commercial trucks and buses.[3]

Figure 2.3 shows a timeline for transitioning to hydrogen fuel cell vehicles. The figure reflects a variety of factors including fuel supply and the time drivers normally own their vehicles. These vehicles include HEV - Hybrid Electric Vehicle, PHEV - Plug-In Hybrid Electric Vehicle, FECV - Fuel Cell Electric Vehicle, FPBEV – Full Performance Battery Electric Vehicle, H2ICV – Hydrogen Internal Combustion Vehicle, CEV – City Battery Electric Vehicle, NEV – Neighborhood Battery Electric Vehicle, FCAPUV – Fuel Cell Auxiliary Power Unit Vehicles.

Figure 2.4 shows the current mix of U.S. fuels used to power vehicles. At present the United States uses about 140 billion gasoline-equivalent gallons. Of this total, only 41,000 gasoline-equivalent gallons are derived from hydrogen.

---

[3] Totals are from 2005 data compiled by Federal Highway Administration and Federal Motor Carrier Safety Administration

Figure 2.3: 2007 Zero Emission Vehicle Panel's Vehicle Projections

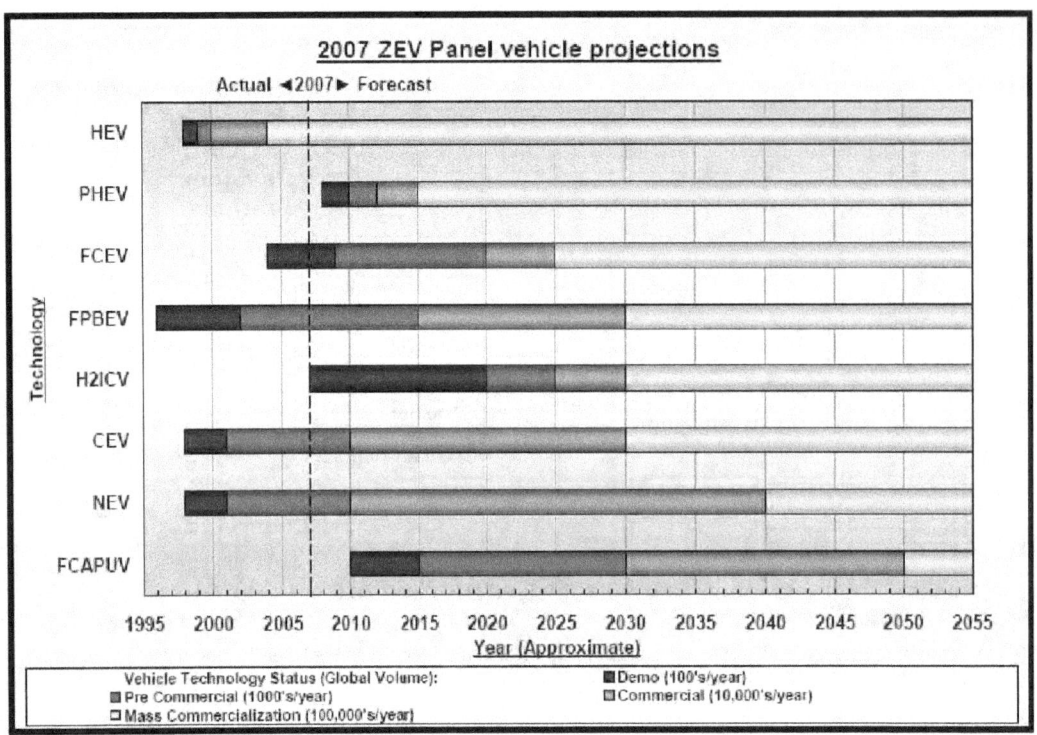

Source: National Research Council. 2008. *Transitions to Alternative Transportation Technologies—A Focus on Hydrogen*, p. 40. The National Academies Press, Washington, DC.

Figure 2.4: Estimated U.S. Consumption of Vehicle Fuels, by Fuel Type, 2003-2006
(in thousand gasoline equivalent gallons)

Fuel Type	2003	2004	2005	2006
Alternative Fuels				
Compressed Natural Gas (CNG)	133,222	158,903	166,878	172,011
Electricity	5,141	5,269	5,219	5,104
Ethanol, 85 percent (E85)[a]	26,376	31,581	38,074	44,041
Hydrogen	2	8	25	41
Liquefied Natural Gas (LNG)	13,503	20,888	22,409	23,474
Liquefied Petroleum Gas (LPG)	224,697	211,883	188,171	173,130
Other Fuels[b]	0	0	2	2
Subtotal	402,941	428,532	420,778	417,803
Biodiesel[c]	18,220	28,244	91,649	260,606
Oxygenates				
Methyl Tertiary Butyl Ether (MTBE)[d]	2,368,400	1,877,300	1,654,500	435,000
Ethanol in Gasohol	1,919,572	2,414,167	2,756,663	3,729,168
Total Alternative and Replacement Fuels[e]	**4,709,133**	**4,748,243**	**4,923,590**	**4,842,577**
Traditional Fuels				
Gasoline[f]	135,330,000	138,283,000	138,723,000	140,146,000
Diesel[f]	41,965,000	41,987,000	43,042,000	44,247,000
Total Fuel Consumption[g]	**177,697,941**	**180,698,532**	**182,185,778**	**184,810,803**

[a] The remaining portion of 85-percent ethanol is gasoline. Consumption data include the gasoline portion of the fuel.

[b] May include P-Series fuel or any other fuel designated by the Secretary of Energy as an alternative fuel in accordance with the Energy Policy Act of 1995.

[c] Estimates for 2003, 2004, and 2005 are revised.

[d] Includes a very small amount of other ethers, primarily Tertiary Amyl Methyl Ether (TAME) and Ethyl Tertiary Butyl Ether (ETBE). Values are rounded to the nearest 100,000 gasoline-equivalent gallons.

[e] A replacement fuel is the portion of any motor fuel that is methanol, ethanol, or other alcohols, natural gas, liquefied petroleum gases, hydrogen, coal-derived liquid fuels, electricity (including electricity from solar energy), ethers, biodiesel, or any other fuel the Secretary of Energy determines, by rule, is substantially not petroleum and would yield substantial energy security benefits and substantial environmental benefits.

[f] Gasoline consumption includes ethanol in gasohol and MTBE. Diesel includes biodiesel. Gasoline and diesel values are rounded to the nearest million gasoline-equivalent gallons.

Notes: Fuel quantities are expressed in a common base unit of gasoline-equivalent gallons to allow comparisons of different fuel types. Gasoline-equivalent gallons do not represent gasoline displacement. The estimated consumption of neat methanol (M100), 85-percent methanol (M85), and 95-percent ethanol (E95) is zero for all years included in this table. Therefore, those fuels are not shown. Totals may not equal sum of components due to independent rounding.

Original Sources: Alternative Fuel Consumption: Energy Information Administration, Office of Coal, Nuclear, Electric and Alternate Fuels. Biodiesel Consumption: Energy Information Administration, Office of Integrated Analysis and Forecasting and U.S. Census Bureau. Ethanol Consumption: Energy Information Administration, Monthly Energy Review, (February 2008). MTBE Consumption: Energy Information Administration, Petroleum Navigator, extracted February 2008. Traditional Fuel Consumption: Energy Information Administration, Petroleum Supply Annual, Volume 1 (September 2007). Highway use of gasoline was estimated as 98.8 percent of consumption, based on data in the Transportation Energy Data Book: Edition 26, prepared by Oak Ridge National Laboratory for the U.S. Department of Energy (May 2007). Diesel consumption was adjusted for highway use by multiplying by .61 derived from Energy Information Administration, Fuel Oil and Kerosene Sales 2005 (December 2007). Diesel consumption was converted to gasoline-equivalent-gallons using heating values from the Energy Information Administration, Monthly Energy Review, (February 2008), Appendix A.

Source: U.S. Department of Energy's Energy Information Administration, http://www.eia.doe.gov/cneaf/alternate/page/atftables/afvtransfuel_II.html#consumption.

According to the NRC, hydrogen fuel cell vehicles are not likely to be cost-competitive until after 2020 when, in a very optimistic scenario, they could comprise about 2 million of the nation's 280 million light-duty vehicles. In that scenario, the number of these vehicles could grow rapidly thereafter to about 25 million by 2030, it added, and by 2050 hydrogen vehicles could account for more than 80 percent of new vehicles entering the fleet. Assuming conventional rates of car buying continue[4], it could take another decade or more to complete the transition. The extent of Federal and private contributions needed to bring the industry to maturity is discussed later in this section.

To help popularize hydrogen vehicles, the NRC recommends consideration of Federal incentives to bridge the cost gap[5] between HFCV and traditional vehicles.[6] "Sustained, substantial and aggressive energy security and environmental policy interventions will be needed to ensure marketplace success for oil-saving and greenhouse-gas-reducing technologies, including hydrogen fuel cell vehicles."[7]

Just as fuel cell performance needs to evolve to make hydrogen fuel cell vehicles an important tool in controlling greenhouse gases, the source of that hydrogen also needs to undergo technological innovation. Today, most hydrogen is produced from natural gas via steam methane reforming. This technology has limited impact on reducing greenhouse gases and improving the environment. Advocates see it as a first step in transition to a hydrogen economy. Coal and nuclear are expected to have the largest positive environmental impact between 2015 and 2030, with newer technologies contributing to a greener environment after 2030. This evolutionary process is shown in figure 2.5.

To realize the vision of creating hydrogen fuel that produces only heat and water, fuel production methods need to change substantially. DOE is funding research and technologies to produce hydrogen from electricity, nuclear energy and clean coal, including building and operating a zero emissions, high-efficiency co-production power plant that will produce hydrogen from coal along with electricity. Nuclear research includes high-temperature thermochemical cycles, high-temperature electrolysis, and reactor/process interface issues.

Figure 2.5: Stationary Power and the Transportation System

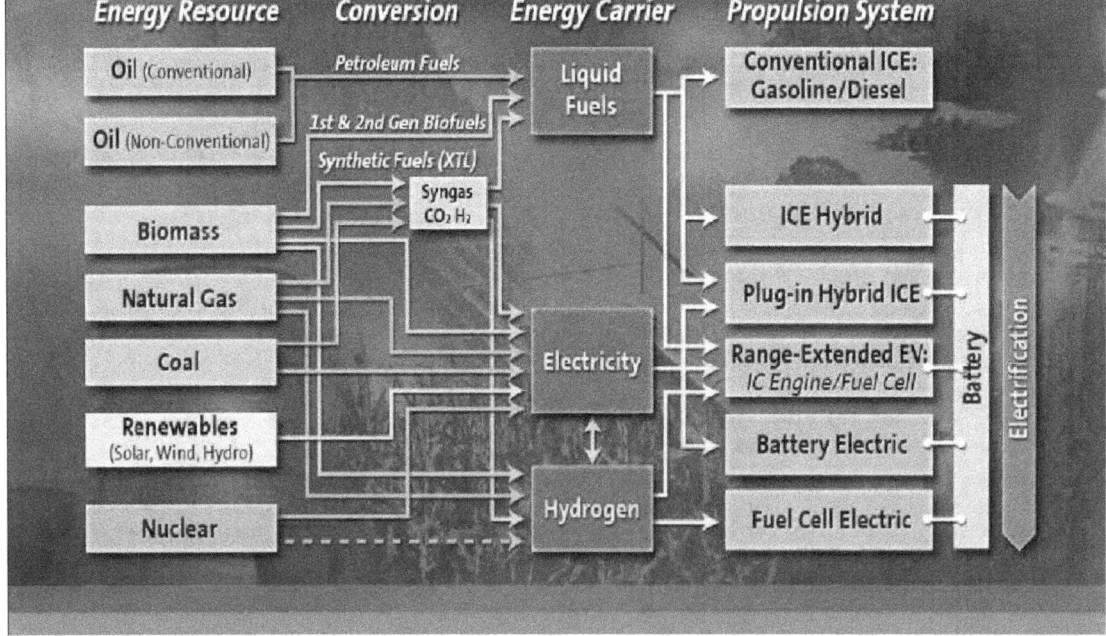

Source: National Research Council. 2008. *Transitions to Alternative Transportation Technologies—A Focus on Hydrogen*, p. 65. The National Academies Press, Washington, DC.

---

[4] Automotive News Data Center reports that there were 7,884,601 cars sold in 2007. There were 8,269,351 trucks and SUVs sold, making for a total of 16,153,952 new vehicles sold in 2007.

[5] HFCVs cost about $7,000 more to produce than their traditionally fueled counterparts.

[6] DOE and Original Equipment Manufacturers are testing 140 hydrogen fueled vehicles. Retail production of these vehicles could begin 2012 -2015.

[7] National Academies of Science, National Research Council, Summit on America's Energy Future: Summary of a Meeting and Transitions to Alternative Transportation Technologies – A Focus on Hydrogen

The ideal end-state for hydrogen focuses on developing advanced technologies from domestic renewable energy resources that minimize environmental impacts. Key DOE research areas include electrolysis, thermochemical conversion of biomass, photolytic and fermentative micro-organism systems, photoelectrochemical systems, and high-temperature chemical cycle water splitting.

However, uncertainty about how, where and with what technologies hydrogen will be produced necessarily creates ambiguities in developing the infrastructure to support hydrogen transport. If, for example, the primary source of hydrogen is natural gas, then very large volumes of natural gas will be required to convert an appreciable fraction of transportation energy consumption to hydrogen. Beyond the relatively straightforward issue of feedstock availability and cost, a natural gas-based hydrogen infrastructure could take several different forms:

► Local, small-scale reforming of hydrogen at or close to service stations. In this case, hydrogen pipeline requirements are modest. Natural gas is piped to service stations, largely through existing infrastructure, and converted to hydrogen at the station. The requirement for a new national hydrogen pipeline infrastructure largely goes away, but service stations become much larger and more expensive. Reforming costs will be high, and it may be difficult to capture the waste heat from reforming for any useful purpose. The inescapable carbon dioxide by-product probably would be vented to the atmosphere.

► Regional reforming of hydrogen from natural gas or, in some cases, coal by plants near major consuming centers. In this case, there is no large-scale interstate hydrogen pipeline transmission system required. Hydrogen may be distributed to service stations by truck or by local distribution pipelines. Reforming costs will be lower, and waste heat can be captured for power generation or other purposes. However, carbon capture and storage will be problematic for consuming centers that are not conveniently located for carbon dioxide use in enhanced oil recovery or geologic sequestration. Most (but not all) consuming centers will be inconveniently located.

► Large-scale central reforming of hydrogen based on nuclear power, coal, or natural gas. In this case, interstate hydrogen transmission infrastructure will be required. Local distribution can be undertaken by truck from pipeline terminals, or by small hydrogen distribution pipelines. On the other hand, central plants can be located by design – where carbon dioxide sequestration is feasible. Additional carbon dioxide pipeline infrastructure will probably be necessary.

It is not clear, at this point, which model of hydrogen production and distribution is most desirable. For transportation planners, the challenge is that the infrastructure requirements for the various models are vastly different.

The challenge for distribution models is that enabling technologies for each scenario requires different additional research, development, testing and funding to reach maturity. NRC estimates that a $40 billion Federal investment would be needed between 2010 and 2030 to enable hydrogen fuel cell vehicles to have the potential to achieve commercial success. NRC expects private sector investment in a successful scenario would have to exceed $100 billion.[8]

Refueling

In addition to the challenges of evolving vehicles and fuels, innovations are needed in how vehicles are refueled. The fundamental quest is to find lighter materials that provide storage rates and refueling times similar to those of fossil fueled vehicles. Figure 2.6 shows the relative volumes needed to travel more than 300 miles.

The energy in 2.2 lb (1 kg) of hydrogen gas is about the same as the energy in 1 gallon of gasoline. A light-duty fuel cell vehicle must store 11-29 lb (5-13 kg) of hydrogen to enable an adequate driving range of 300 miles or more. Because hydrogen has a low volumetric energy density (a small amount of energy by volume compared with fuels such as gasoline), storing this much hydrogen on a vehicle using currently available technology would require a very large tank—larger than the trunk of a typical car. Advanced technologies are needed to reduce the required storage space and weight.

---

[8] National Academies of Science, National Research Council, Summit on America's Energy Future: Summary of a Meeting and Transitions to Alternative Transportation Technologies – A Focus on Hydrogen

Storage technologies under development include high-pressure tanks with gaseous hydrogen compressed at up to 10,000 pounds per square inch, cryogenic liquid hydrogen cooled to -423°F (-253°C) in insulated tanks, and chemical bonding of hydrogen with another material (such as metal hydrides).

DOE and DOT are funding research to explore fuel cell bus operations in 14 cities and some of these conveyances will be fueled with hydrogen. They also are studying how innovative composites can permit higher rates of hydrogen storage. The goal is to identify materials that increase the amount of hydrogen a tank holds and facilitate its flow during refueling.

The infrastructure requirements for different vehicle onboard storage designs differ. If the preferred vehicle design is for liquefied hydrogen, this presents a significant challenge for the design of the infrastructure. Gaseous hydrogen has to be cooled, compressed, and stored before delivery to the vehicle at yet to be defined locations conditions and locations. An all-liquefied hydrogen supply chain would be difficult and is probably infeasible. Safety and cost considerations will probably argue for liquefaction relatively close to the point of sale, but probably not at the service station (though this is possible). However, a liquefied hydrogen service station is a much more elaborate facility than a gaseous hydrogen station, and may present somewhat different safety and siting considerations.

Raising storage pressures to 10,000 psi would likely increase the attractiveness of hydrogen vehicles. However, 10,000 psi storage raises design questions for service stations and local distribution. The fundamental question is whether service stations locally compress hydrogen to 10,000 psi for delivery to vehicles, or whether some portion of the supply chain (delivery trucks, for example) ought to operate at higher pressures as well.

As in other cases, the challenge for infrastructure planners is to design an infrastructure when key parameters remain uncertain.

Figure 2.6: Relative Volume Needed for Hydrogen Storage to Achieve a Range of More Than 300 Miles

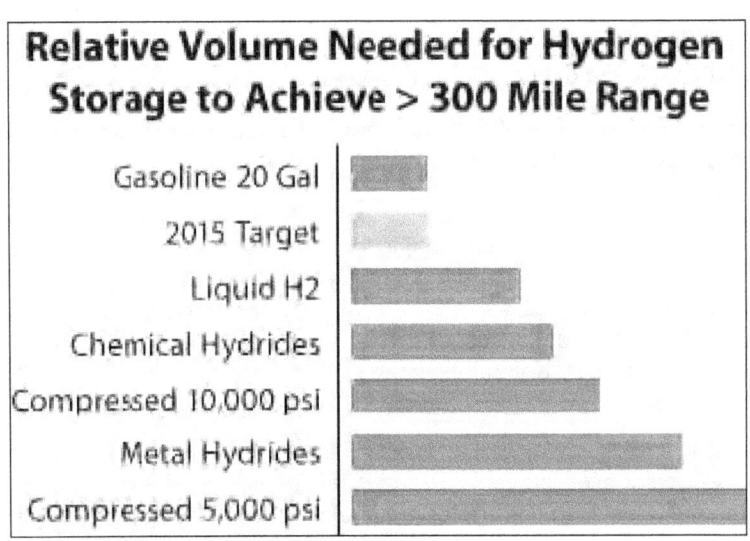

Source: U.S. Department of Energy, Office of Energy Efficiency and Renewable Energy, Hydrogen Learning Demonstration.

According to DOE, its Hydrogen Learning Demonstration has collected data on refueling rates, which has shown that, on the average, at 350 bar pressure, the refueling rate is 0.81 kg/min (with over 25% at a rate of more than 1 kg/min). At 700 bar pressure, the average refueling rate is 0.59 kg/min (with only 3% over 1 kg/min).

PUBLIC AND PRIVATE SECTOR ORGANIZATIONAL CHALLENGES

The organizational challenges facing government and others fall into three major components – coordination of Federal, State and local governments; land use and site planning; and public perception and education.

Governmental Coordination

NRC notes that one of the most important challenges in a transition from petroleum fuels is a consistent and clear framework of Federal, State and local requirements for the storage and use of alternative fuels. This is especially critical for hydrogen since it requires a separate infrastructure from that used for today's petroleum vehicles.

Because Congress adopted the United Nation's Dangerous Goods Code to govern all hazardous materials moving in U.S. commerce, there is consistency in the requirements governing the transport of these fuels. Federal law requires that only DOT's Pipeline and Hazardous Materials Safety Administration (PHMSA) may grant any deviations or waivers from these international rules.

NRC and others identify standardization of requirements and/or production processes as key to facilitating the widespread adoption of any innovation. The ability to quickly build and deploy hydrogen processing plants and fueling stations across the nation requires that manufacturers have certainty in the products they design and develop for distribution across the nation. Consumers require the certainty that standardization will bring to buying, using, repairing and fueling hydrogen vehicles.

Government, especially the Federal Government, can play a vital role in providing this certainty either through nationwide regulation or strong support of national industry standards. As noted in the *Safety Codes and Standards* section of this report, DOE, DOT and the other Federal hydrogen agencies are working with the key national standards bodies to:

▶ Identify and publicize existing standards

▶ Identify areas where standards are lacking

▶ Support the research that develops the data for these new standards

▶ Help build broad acceptance of these data-driven requirements

▶ Support their incorporation into new codes and standards.

Land Use & Planning, Siting

The planning process for the creation of infrastructure – whether a fueling station or a stretch of highway – is controlled at the State and/or local levels of government. While the Federal Government may issue guidelines for these activities, there is little Federal control over what are basically local land-use decisions.

Although land use authorities are familiar with the requirements for creating a safe and efficient fossil fuel station, they often are unaware of what exactly is needed for hydrogen fueling station. As DOE has found, this lack of familiarity creates reluctance to approve facilities and makes securing approval longer and more costly. In some cases, local laws would prohibit or deter creation of alternative fuel stations.

Land use and planning are especially critical issues for the building of retail hydrogen refueling facilities. Unlike some other alternative fuels, hydrogen pumps cannot simply be added to existing fossil fueling stations. Because of safety and inherent handling properties, hydrogen refueling stations require separate infrastructure to meet these different handling requirements.

As noted by NRC, a full-size hydrogen refueling unit added to a conventional fueling station with a minimart would an additional 7,200 square feet of space. This would bring the footprint to almost 14,000 square feet (7,200 + 6,500). Even if a smaller (e.g., 100 kg/d) hydrogen fueling unit is used, a station would still require about 2,200 additional square feet. In urban areas, this footprint could limit the number of existing sites that could be used for both purposes. It also opens up the possibility that many of the hydrogen refueling sites will be at nontraditional locations such as shopping malls and big-box retailer parking areas or even auto dealerships.[9]

Public Perception and Education

If the nation is to lessen its dependence on fossil fuel, then the public, as well as State and local decisionmakers, need to better understand hydrogen, hydrogen infrastructure and the social benefits of making alternative fuel investments. This information is crucial to allaying public concern and opposition to any kind of development.

It will take a large-scale, concerted effort to help overcome this reluctance to invest and build, especially when cutting edge technologies are involved. State and local planners and officials, as well as private sector decisionmakers, will need training, as well as opportunities to collaborate about making safe and smart investments. The public will need similar opportunities to better understand the societal benefits of these investments. Future Federal programs will need to integrate these considerations.

COMMERCIAL SECTOR

Just as the public sector faces coordination and land use and planning issues, the private sector has similar concerns. These concerns encompass station start-up; network maturity, station standardization, fuel quality and quantity; and insurance and liability. Federal guidance, as well as education and outreach could create the innovative partnerships needed in the transition from a fossil fueled economy.

Station Start-Up

For infrastructure and its supporting land use and planning activities, an important consideration is whether there will be a transition from portable to permanent stations. Fossil fuels in the late 19th century and very early 20th century were sold at pharmacies. Users then received regular deliveries at home or at their places of business. It took about 15 years from the time the first car was made in the United States for the first public fueling station to open.

Investment strategies for alternative fueling stations including hydrogen have similar variation in cost and level of investment. As shown in a University of California – Davis study, A Near-term Economic Analysis of Hydrogen Fueling Stations (Jonathan Weinert, Dr. Joan Ogden), the costs for starting a station can vary greatly based on the type created.[10]

To facilitate private sector investment in portable and permanent refueling, there needs to be clear public sector direction on the preferred migration strategy as well as on the types of equipment and land-use configurations needed to get there. As mentioned elsewhere in this report, the widespread adoption of national model codes and requirements appear essential to meeting this challenge.

Network Maturity

One of the special challenges facing a hydrogen infrastructure is at the retail level. At the bulk level, there are 700 miles of DOT-regulated transmission pipeline and, according to EIA, about 500 miles of distribution pipeline dedicated to hydrogen movements. The fuel also is able to share the liquid natural gas network. However, these benefits are stymied at the retail level, where there are about 60 vehicle refueling stations, most of which are located in California. Network maturity also affects other alternative fuels as outlined in Appendix B.

A key unanswered challenge is how hydrogen distribution and retail networks will grow. As NRC noted, there is debate as to whether specific markets or regions, such as California, should be targeted for hydrogen investment or if a national strategy of a station every 25 miles should be pursued. Coordination of fuel supply with vehicle distribution will be an important area of public/private cooperation.[11]

---

[9] National Research Council, *Transition to Alternative Transportation Technologies – A Focus on Hydrogen*, 2008.
[10] http://pubs.its.ucdavis.edu/publication_detail.php?id=46
[11] *Transitions to Alternative Transportation Technologies – A Focus on Hydrogen*.

Station Standardization

Another concern for investors will be standardization of requirements for planning, constructing and operating refueling stations. The use of model national codes for planning and constructing facilities helps investors leverage the lessons learned from earlier activities so that subsequent efforts can be delivered more quickly and cost effectively. Standardization encompasses issues from the performance of pumps and storage tanks to the proper siting of this equipment at stations.

Improving the compatibility and interoperability of station equipment is another way to facilitate deployment because it reduces deployment costs as well as the time needed to build this infrastructure.

Fuel Quality

Whether fueling their vehicles with gasoline, biofuels or hydrogen, consumers want certainty about the quality of the product they are buying. An important component of the infrastructure maturation process is the development of fuel quality standards and their widespread adoption and implementation by industry.

At present, DOE, International Standards Organization (ISO), the Society of Automotive Engineers (SAE), the California Fuel Cell Partnership (CaFCP), and the New Energy and Industrial Technology Development Organization (NEDO)/Japan Automobile Research Institute (JARI) are working to put in place these standards for hydrogen fuel cells.

This partnership of domestic and international interests has two goals. The first is to identify and exclude potential contaminants from the automotive fuel cell or in on-board hydrogen storage systems. The second is to balance the extremely high cost of providing extremely pure hydrogen with the life-cycle costs of the complete hydrogen fuel cell vehicle "system." Currently, partnership researchers are working together to assess the influence of different contaminants and their concentrations to develop a process whereby the hydrogen quality requirements may be determined and broadly adopted. Their success will determine consumer acceptance of hydrogen fuel cells.

Insurance/Liability

Insurance rates are based on risk and potential liability to the insurer. Insurers conduct formal assessment of financial risk (e.g., likelihood and cost of adverse events) based on historical records of frequency and cost impact of an adverse event. The industry then establishes premiums to cover this risk. Insurers generally charge high fees, limit coverage and/or require high deductibles for covering extraordinary situations where historical experience is thin. The burgeoning use of fuel cells for vehicles and conveyances and the development of stations to refuel them is a prime example of such a situation.

Until a record of successful hydrogen station operation is established, insurance and liability requirements could deter many potential investors from financing hydrogen refueling stations. It will take Federal, State and local officials working together with the insurance industry and station investors and operators to overcome this significant practical barrier.

One public/private partnership recommended the following strategies to address the insurance issue:[12]

▶ facilitating the collection of statistics and analysis underwriters need to offer rates commensurate with the true risk factors associated with the use of hydrogen in transportation applications;

▶ requiring plans for hydrogen fueling stations to include risk assessment and risk management elements in their permitting submittals;

▶ creating state insurance pools for partial coverage of deductibles;

▶ temporarily limiting liability for adverse events where permit requirements (such as maintenance, training, and inspections) were followed; and,

▶ allowing station installers/operators to self-insure and setting liability limits.

---

[12] California 2010 Hydrogen Highway Network, *Implementation Topic Team Report: Codes & Standards, Insurance & Liability*, Jan. 5, 2005

SAFETY AND EMERGENCY RESPONSE CHALLENGES

Any transition to hydrogen-fueled transportation will require the creation and/or updating of safety codes and standards for the safe handling of the fuel during manufacture, when in transit or at refueling stations. It also will require additional training and tools for emergency responders because of the differences in handling properties between hydrogen and fossil fuels.

Safety Codes & Standards

Although the process of fueling a hydrogen car is not much different than refueling a gasoline vehicle, hydrogen needs to be handled differently from petroleum fuels. As a result, there is a need for hydrogen specific codes and standards for storage, fueling and emergency response. Table 2.1 outlines the objectives of codes and standards and how each are used.

As with any fueling station, hydrogen stations typically combine bulk storage and dispensing. They may provide gaseous hydrogen, liquid hydrogen, or both to cars, buses, or other vehicles such as forklifts. Like other fuels, hydrogen stations can be on private property or industrial grounds, as well as part of retail fueling stations that also provide gasoline, diesel, or other fuels.

Table 2.1. Objectives, Codes and Standards

	Objectives, Codes and Standards
Objectives	▶ Provide information needed to safely build, maintain, and operate equipment, systems, and facilities ▶ Help ensure uniformity of safety requirements ▶ Give local inspectors and safety officials the information needed to approve systems and installations
Codes	▶ Guide the design of the built environment ▶ Are adopted by local jurisdictions ▶ Are used to refer to or invoke standards for equipment used within a built environment
Standards	▶ Are adopted by local jurisdictions ▶ Are used to refer to or invoke standards for equipment used within a built environment ▶ Define rules, guidelines, conditions, or characteristics for products or related processes ▶ Apply generally to equipment or components ▶ Can have regulatory-like status when cited in codes or other regulations

Source: U.S. Department of Energy, http://hydrogen.pnl.gov/FirstResponders/Flash/Controller.faces

Hydrogen stations are designed with a number of sensors and safety systems to protect against potential hazards. Sensors detect leaks and computers monitor all operating systems to ensure against problems. Flame detectors watch the refueling station at all times.

Hydrogen fires normally are not extinguished until the supply of hydrogen has been shut off due to the danger of re-ignition and explosion. Personnel who work around hydrogen should be trained in the characteristics of hydrogen fires and proper procedures for dealing with them. For example, a hydrogen fire is often difficult to detect without a thermal imaging camera or flame detector. Emergency responders need to let a gaseous hydrogen fire burn, but spray water on adjacent equipment to cool it.

Because of these differences between traditional petroleum-based fuels and hydrogen, safety codes and standards repeatedly have been identified as a major institutional barrier to deploying hydrogen technologies and developing a hydrogen economy. To enable the commercialization of hydrogen in consumer products, new model building codes and equipment and other technical standards need to be developed and recognized by Federal, State, and local governments.

DOE, DOT and other Federal partners are working to identify needed codes and standards, facilitate their development with the pertinent stakeholders and support publicly available research and certification investigations necessary to provide the data and science for promulgating them.

A large number of possible codes and standards can come into play for permitting design and construction of hydrogen fueling stations. Additional Federal, State, and local requirements also may apply. These codes are needed to ensure the safety of employees and customers. They include the proper design, location, and operation of storage and dispensing equipment and the proper installation and operation of leak detection, fire detection, and fire suppression equipment. In addition, incompatible materials or improperly installed equipment can lead to fuel contamination, which can degrade the performance of the fuel cells that power hydrogen-fueled vehicles.

With respect to fueling stations and fuel cell installations, DOE has worked with the National Fire Protection Association to develop the necessary hydrogen codes. An update to current standards is in progress and should be ready by the end of the calendar year. This comprehensive update is aimed at standardizing and speeding up the permitting process.[13] In addition, DOE held 15 workshops across the country to educate more than 250 code officials on the permitting process for hydrogen fueling stations.

Continued development and updating of standards, as well as education of officials on the need to adopt and implement them at all levels of government, will be an on-going challenge in planning, building and deploying the network of production and dispensing facilities needed to make hydrogen-based transportation a reality.

A list of the key standards and code organizations include:

- American Gas Association (AGA)
- American Petroleum Institute (API)
- American Society for Testing and Materials (ASTM)
- American Society of Heating, Refrigerating and Air-Conditioning Engineers (ASHRAE)
- American Society of Mechanical Engineers (ASME)
- Canadian Standards Association (CSA)
- Compressed Gas Association (CGA)
- Consumer Product Safety Commission (CPSC)
- Environmental Protection Agency (EPA)
- Gas Technology Institute (GTI)
- Institute of Electrical and Electronics Engineers (IEEE)
- International Code Council (ICC)
- National Fire Protection Association (NFPA)
- National Highway Traffic Safety Administration (NHTSA) (USDOT)
- National Institute of Standards and Technology (NIST) (USDOC)
- Occupational Safety and Health Administration (OSHA) (USDOL)
- Pipeline and Hazardous Materials Safety Administration (PHMSA) (USDOT)
  Society of Automotive Engineers (SAE)
  Underwriters Laboratory (UL)

Regardless of the alternative fuel, the transition from gasoline and diesel will require government to act as convener and facilitator. Congress will need to support basic and applied research, as well as outreach to stakeholders and public sector participation in standard setting bodies. It also will need to provide the Federal agencies with the technical resources to harmonize development of domestic standards with interna-

---

[13] http://www.nfpa.org/assets/files/PDF/CodesStandards/HCGNASAHydrogenStd.pdf;
http://www.nfpa.org/assets/files/PDF/CodesStandards/HCGNASAHydrogenStd.pdf.

tional standards and help resolve conflicts in international requirements, i.e. International Organization for Standardization (ISO), International Electrotechnical Commission (IEC), and Working Party on Pollution and Energy (GRPE). Appendix C provides information on DOE's international efforts.

Among the challenges government faces are:

- Limited Government Influence on Model Codes
- Limited State Funds for Adoption of New Codes
- Large Number of Local Government Jurisdictions Needed to Adopt these Standards.
- Lack of Consistency in Training of Officials
- Limited US Role in the Development of International Standards including Inadequate Representation at International Forums
- International Competitiveness and the Resulting Conflicts between Domestic and International Standards
- Lack of National Consensus on Appropriate Codes and Standards Requirements
- Jurisdictional Legacy Issues
- Insufficient Technical Data to Revise Standards

Emergency Response

Hydrogen has been delivered safely for decades, mostly by pipeline or over the road. The current U.S. hydrogen pipeline infrastructure is small - about 700 miles of DOT-regulated transmission lines, compared to more than a million miles of DOT-regulated natural gas transmission pipeline - so hydrogen is often delivered by trucks carrying gaseous or liquid hydrogen in cylinders or tanks.

DOT's Pipeline and Hazardous Materials Administration (PHMSA) administers and enforces Title 49, Code of Federal Regulations requirements for the transport and storage of hydrogen, along with other fuels and hazardous materials. This includes specifying approved shipping containers, including pipelines, defining testing, maintenance, and inspection requirements for safe transport and handling. Aluminum or steel cylinders are a common approved packaging for compressed gases, including hydrogen.

Tube trailers transport bulk quantities of hydrogen gas, while cargo tanks carry bulk liquid hydrogen. Placards and material identification numbers are required to be displayed on bulk transport vehicles to help first responders recognize the material and respond appropriately in the event of an emergency.

Because of the differences in the handling properties of hydrogen and petroleum based fuels, a suitably trained emergency response force is an essential component of a viable infrastructure. Training of emergency response personnel is a high priority, not only because these personnel need to understand how to deal with a hydrogen-related emergency situation, but also because firefighters and other emergency workers are influential in their communities and can be a positive force in the introduction of hydrogen and fuel cells into local markets.

DOE and DOT, working the Occupational Safety and Health Administration (OSHA) and National Fire Protection Association (NFPA), are developing frameworks for hazardous materials emergency response training, and a tiered hydrogen safety education program for emergency responders. In 2007, the first training tools were released. They provided a basic awareness about hydrogen and a high level overview of how to handle these commodities.[14] More sophisticated and rigorous materials are in development.

SUSTAINED COMMITMENT CHALLENGES

Taking an enterprise view of the transition to cleaner fuels for transportation will involve a long term focus by the Federal Government and its private sector counterparts.

---

[14] http://hydrogen.pnl.gov/FirstResponders.

Long term areas of effort[15] could include:

▶ Price supports/incentives/tax credits to offset the higher cost of buying an alternatively fueled vehicle rather than one powered by gasoline;

▶ Grants and/or loans to create the network of refueling stations;

▶ Grants for basic and applied research to develop fuel cell batteries with operating lives of at least 5,000 hours (equivalent to a 300-mile traditional trip);

▶ Grants for basic and applied research, testing and deployments to develop the hydrogen generating capacity to fuel at least 220 million vehicles. Today in the U.S there are 347 million registered vehicles.[16]

Table 2.2 captures key areas of investment for a large scale transition to hydrogen fuel cell vehicles.

Table 2.2. Key Investment Areas

TABLE S.1 Summary of Cumulative Budget Roadmap Costs for Transition to Hydrogen Fuel Cell Vehicles (maximum practicable number of vehicles by 2020)

Cost Elements	Total Cumulative Cost, 2008-2023	Average Cost per HFCV on Road 2008-2023ᵃ
"Base vehicle" cost of conventional vehicles	$128 billion	$23,000
Average incremental fuel cell vehicle cost relative to conventional gasoline vehicles	$40 billion	$7,000ᵇ
Total purchase cost of fuel cell vehicles	$168 billion	$30,000ᶜ
Infrastructure capital cost for hydrogen supply	$8 billion	$1,500
Total operating cost for hydrogen supply	$8 billion	$1,500
Total cost of hydrogen supply	$16 billion	$3,000
Total cost for vehicles and hydrogen fuel supply	$184 billion	$33,000
Estimated government share of total vehicle and hydrogen fuel supply cost	$50 billion	$8,500
Government RD&D funding	$5 billion	$1,000
Private RD&D funding	$11 billion	$2,000
Total funding for government and private RD&D	$16 billion	$3,000
Total cost for vehicles, hydrogen, and all RD&D	$200 billionᵈ	$36,000
Estimated government share of total cost for vehicles, hydrogen, and RD&D	$55 billion	$9,500

ᵃRounded estimates based on 5.54 million HFCVs on the road in 2023.
ᵇThe final (learned-out) incremental cost per vehicle in 2023 is $3,600.
ᶜThe final (learned-out) cost per vehicle in 2023 is $27,000.
ᵈIncludes $128 billion "base vehicle" cost of conventional vehicles that would have been purchased instead of HFCVs.

NOTE: All costs in constant 2005 U.S. dollars.

Source: Source: Hydrogen Energy Center, *The Future of Hydrogen: an Alternative Transportation Analysis for the 21st Century*, National Hydrogen Association Webinar, Oct. 23, 2008.

NRC estimated the Federal Government's contributions as being roughly $55 billion from 2008 to 2023 (when fuel cell vehicles would become competitive with gasoline-powered vehicles). This funding includes a substantial R&D program ($5 billion), support for the demonstration and deployment of the vehicles while they are more expensive than conventional vehicles ($40 billion), and support for the production of hydrogen ($10 billion). Private industry, it added, would be investing far more, about $145 billion for R&D, vehicle manufacturing, and hydrogen infrastructure over the same period.

NRC further refined this estimate in its Summit on America's Energy Future: Summary of a Meeting. There it notes that the private sector cost for hydrogen infrastructure would be about $400 billion by 2050

---
[15] Recommendations reflect discussions and proposals included in NRC's Transitions to Alternative Transportation Technologies – A Focus on Hydrogen, and its Summit on America's Energy Future: Summary of a Meeting, California Fuel Cell Partnership – Hydrogen Fuel Cell Vehicle and Station Deployment Plan and: A Strategy for Meeting the Challenge Ahead Action Plan and National Hydrogen Association – The Future of Hydrogen: An Alternative Transportation Analysis for the 21st Century.
[16] Federal Highway Administration and Federal Motor Carriers Safety Administration.

to support 220 million vehicles. This total also would include 180,000 stations, 210 central plants, and 80,000 miles of pipeline.

In a 2007 paper authored by General Motors Research & Development Center and Shell Hydrogen, the private sector researchers estimate that the cost to construct 12,000 refueling stations is between $10B and $15B. This network would put 70 percent of the U.S. population, now living in the nation's 100 largest cities, within a two-mile radius of a refueling station and connect these cities with a station every 25 miles along the interstate highway system. The authors provided no timelines for funding and deploying these investments.

In its 2009 Action Plan, the California Fuel Cell Partnership estimates its station startup costs at about $3 million to $4 million each. These estimates are shown in Table 2.3.

Table 2.3. Costs of Hydrogen Stations in California Through 2012

	2009	2010	2011	2012
New stations funded	10	9	11	10
Total station costs (millions)	$30.8	$37.5	$51.2	$59.5
Cumulative total station costs (millions)	$30.8	$68.3	$119.5	$179.0
Gov't cost share for stations (millions)	$23.4	$27.2	$35.9	$31.1
Regulatory development and outreach costs (millions)	$1.0	$0.5	$0.5	$0.5
Cumulative gov't cost (millions)	$24.4	$52.1	$88.5	$120.1
Total cumulative costs (millions)	$31.8	$69.8	$121.5	$181.5

Source: California Fuel Cell Partnership, *Hydrogen Fuel Cell Vehicle and Station Deployment Plan: A Strategy for Meeting the Challenge Ahead, Action Plan*, February 2009, p. ii.

In summary, as shown in Table 2.4, the challenges of hydrogen infrastructure fall into the following key areas:

▶ Technology

▶ Public and Private Sector Organizations

▶ Commercial Sector Issues

▶ Safety Codes and Standards including Emergency Response

▶ Sustained Commitment

Table 2.4 (pp. 24- 28) and Table 3.1(pp. 32-34) highlight where the nation is on transition to a hydrogen infrastructure as well as the remaining barriers and possible strategies that could facilitate the journey.

Table 2.4. Hydrogen Infrastructure Challenges

Challenge	Status	Remaining Barriers and Strategies
Technology Challenges		
Fuel Cell Performance	The NRC reported fuel cell "stack" operating life has grown from about 1,000 hours in 2004 to more than 1,500 hours. In 2008, DOE's National Renewable Energy Laboratory documented an independent validation of 140 fuel cell vehicles that showed nearly 2,000 hours under real world conditions. More recently, fuel cells have lasted as long as 7,300 hours in laboratory testing. Fuel cells also are beginning to match the fossil-fueled distances between refuelings – about 300 miles	The Federal goal for fuel cell stack life is 5,000 hours, equivalent to 150,000 miles of engine life for gasoline-powered driving. The goal for on-board hydrogen storage systems is $2 per kilowatt hour and $30 kilowatt for fuel cells. NRC reported that Toyota testing did exceed the 300-range goal but at an estimated cost of $15 to $18 per kilowatt hour for on-board storage. The size and weight of current fuel cell and fuel storage systems must be further reduced for commercial success.
Vehicle Supply	DOE has been working with original equipment manufacturers (OEMs) to test and demonstrate 140 hydrogen fuel cell vehicles.	NRC estimates hydrogen fuel cell vehicles are not likely to be cost-competitive until after 2020 where they could comprise about 2 million of the nation's 280 million light-duty vehicles. Federal incentives potentially could bridge the cost gap between HFCV and traditional vehicles. California's mandate for zero emissions vehicles offers an opportunity to increase vehicle supply.
Fuel Supply	DOE has funded research to produce hydrogen from electricity, nuclear energy and clean coal. Key research areas include electrolysis, thermochemical conversion of biomass, photolytic and fermentative micro-organism systems, photoelectrochemical systems, and high-temperature chemical cycle water splitting. Depending on the technology used to produce the hydrogen, the cost per gasoline gallon equivalent (gge) now ranges between $3 and $9. The DOE target is $2 - $3 per gas gallon equivalent.	New hydrogen production technology is needed to increase the output of hydrogen production, reduce its cost, as well as reduce greenhouse gases. These technologies will need additional research, development, testing and funding to bring to maturity.
Refueling	DOE has achieved its milestone of a refueling time of 5 minutes or less for 5 kg of hydrogen at 350 bar dispensing pressure. DOT is funding researchers to explore fuel cell bus operations in 14 cities and some of these conveyances will be fueled with hydrogen. DOE and DOT also are working to discover how innovative composites can permit higher capacities of hydrogen storage.	More research is needed to find lighter materials that can store more hydrogen and have refueling times similar to traditional vehicles. Research also is needed to increase supply and availability. Costs for hydrogen production and bulk storage of fuel at retail stations must be reduced for successful commercialization. Lower cost storage tanks are important for near and mid-term success, while low pressure storage technologies are need for commercial success of all vehicle types. Methods for greater vehicular hydrogen storage capacities within the packaging constraints of a vehicle (e.g. cryocompressed and materials options) still require R&D and refueling approaches must be addressed.
Public & Private Sector Organizational Challenges		
Federal, State & Local Coordination	DOE, DOT, industry, and codes and standards organizations are facilitating and expediting codes and standards development to recognize hydrogen as a fuel gas. Standardized rules for transport & in-transit handling. PHMSA has adopted requirements similar to those contained in the UN Model Regulations into 49 CFR based on a series of rulemakings which harmonize domestic regulations with international requirements.	Today, there are approximately 44,000 jurisdictions with their own planning processes. More work is needed to facilitate common codes and standards.

Land Use & Site Planning	Siting is a State, regional or local decision. DOE launched a web-based permitting compendium to promote standardization of permitting hydrogen and fuel cell installations and conducted workshops to education code officials to improve the permitting process. The Technical Reference on Hydrogen Compatibility of Materials (www.ca.sandia.gov/matlsTechRef); the Regulators' Guide to Permitting Hydrogen Technologies (http://www.pnl.gov/fuelcells/docs/permit-guides/module2_final.pdf); and the Web sites "Hydrogen Safety Best Practices" (http://www.h2bestpractices.org/) and "Permitting Hydrogen Facilities (http://www.hydrogen.energy.gov/permitting/	Land use authorities' lack of awareness of what is needed for a hydrogen fueling station leads to reluctance to approve facilities.
Public Perception & Education	Some work initiated. Broad understanding is key to acceptance. DOE is conducting public outreach to increase knowledge and understanding of hydrogen technologies. Last summer, DOT, DOE and California Fuel Cell Partnership participated in a national media tour of hydrogen vehicles to increase public awareness and acceptance.	State and local planners and officials, as well as private sector decision-makers, need training and opportunities to collaborate about making safe and smart investments. The public will need similar opportunities to better understand the societal benefits of these investments. More work is needed to facilitate conformity. Future Federal programs will need to integrate these considerations in all of its future program requirements as it invests in research and technological innovations.
Commercial Sector Challenges		
Station Start Up	DOE estimates there are about 60 hydrogen fueling stations across the nation.	Investment strategies for alternative fueling stations including hydrogen have similar variation in cost and level of investment. The costs for starting a station can vary greatly based on the type created. In its 2009 Action Plan, the California Fuel Cell Partnership estimates its station startup costs at about $3M - $4M each.
Network Maturity	There are 700 miles of DOT-regulated hydrogen transmission pipeline in the United States Energy Information Administration estimates that there is another 500 miles of distribution pipeline. DOE estimates there are about 60 hydrogen fueling stations across the nation.	Decisions are needed on whether specific markets or regions, such as California, should be targeted for hydrogen investment (DOE) or if a national strategy of a station every 25 miles should be pursued (NRC). Coordination of fuel supply with vehicle distribution will be an important area of public/private cooperation.
Station Standardization	Stations are currently being built with both 350 and 700 bar dispensing pressure. Dialogue will continue with auto manufacturers and fuel providers regarding preferred technology, performance, and cost issues. As part of learning demonstrations, DOE is working on increasing the compat bility of components and standardizing the process. Twenty stations have been successfully opened a result.	Requirements for planning, constructing and operating refueling stations must be standardized. This standardization includes the use of model codes for planning and construction, pump performance standards and proper siting of storage tanks.
Fuel Quality & Quantity	At present, DOE, International Standards Organization (ISO), the Society of Automotive Engineers (SAE), the California Fuel Cell Partnership (CaFCP), and the New Energy and Industrial Technology Development Organization (NEDO)/Japan Automobile Research Institute (JARI) are working to improve hydrogen quality, quantify impacts and mitigate effects.	Identify and exclude potential contaminants on the automotive fuel cell or on-board hydrogen storage systems, as well as assess their impacts and cost/performance trade-offs. Balance extremely high costs of providing extremely pure hydrogen with the life-cycle costs of the complete hydrogen fuel cell vehicle "system."
Insurance & Liability	DOE is developing a strategy to address the issues and concerns associated with insurance for deployment of hydrogen infrastructure.	Insurance and liability requirements could deter many potential investors from financing hydrogen refueling stations. It will take Federal, State and local officials, the insurance industry and station investors and operators working together to over come this barrier.

	Safety Codes & Standards Challenges	
Standards Codes & Permitting	Code framework was created, code officials were trained and hydrogen specific code documents were prepared. DOE and its Federal partners are working to identify needed codes and standards, facilitate their development with the pertinent stakeholders and support publicly available research and certification investigations necessary to provide the data and science for promulgating them.	To enable the commercialization of hydrogen in consumer products, new model building codes and equipment and other technical standards need to be developed and recognized by Federal, State, and local governments. Among the challenges government faces are: Limited government influence on model codes; Limited State funds for adoption of new codes; Large number of local government jurisdictions Needed to adopt these standards; Lack of consistency in training of officials; Limited US role in the development of international standards including inadequate representation at international forums; International competitiveness and the resulting conflicts between domestic and international standards; Lack of national consensus on appropriate codes and standards requirements; Jurisdictional legacy issues; Insufficient technical data to revise standards.
Safety & Emergency Response	Basic trainings have been initiated. An on-line educational package on hydrogen safety information for first responders was published by DOE and has been completed by over 7,000 users to date. DOT's PHMSA administers and enforces Title 49, Code of Federal Regulations requirements for the transport and storage of hydrogen, along with other fuels and hazardous materials. This includes specifying approved shipping containers, including pipelines, have testing, maintenance, and inspection requirements for safe transport and handling. DOE and DOT have developed and are continuing to develop hydrogen safety programs for emergency responders.	More work needed to reach broader audience with better trainings and tools. A suitably trained emergency response force is an essential component of a viable infrastructure. Training of emergency response personnel is a high priority, not only because these personnel need to understand how to deal with a hydrogen-related emergency situation, but also because firefighters and other emergency workers are influential in their communities and can be a positive force in the introduction of hydrogen and fuel cells into local markets.
	Sustained Commitment Challenges	
Federal	The Federal Government already funds R&D programs developing technologies, fuels and fuel supplies. Government policies are being developed to support infrastructure development and to facilitate the development of business cases to support private sector investment. Tax credits and other financing tools support vehicle conversion.	NRC estimated the Federal Government's required contributions as roughly $55 billion from 2008 to 2023 for R&D and support for the demonstration and deployment of the vehicles, which initially are expected to be more expensive than conventional vehicles. NRC supports R&D for hydrogen production and demonstration projects as key to acceptance and broad deployment. These include fuel cell vehicle components and systems, hydrogen production, delivery and storage and safety, codes and standards activities, emergency response, and technology validation.
Marketplace	Hydrogen technologies will need additional research, development, testing and funding to bring to maturity. According to NRC, $40 billion of the $55 billion Federal investment for hydrogen between 2010 and 2030 is to make fuel cell technology more affordable to consumers. Private sector investment (consumer and industry) is expected to be in excess of $140 billion.	Private sector research and innovative approaches to private/public partnerships are key to acceptance and broad deployment. NAS notes that the private sector cost for hydrogen infrastructure would be about $400 billion by 2050 to support 220 million vehicles. This total also would include 180,000 stations, 210 central plants, and 80,000 miles of pipeline.
Source: U.S. Department of Transportation, Research & Innovative Technology Administration and U.S. Department of Energy, Office of Efficiency and Renewable Energy's Fuel Cell Technologies Program.		

Federal Progress to Date

Table 3.1 provides additional information on the technical aspects of progress to date.

There are many private and public sector interests working to facilitate the nation's transition to a hydrogen fueled economy. Within the Federal Government, the lead agency is the Department of Energy. The Departments of Transportation (DOT) and Agriculture (USDA), the Environmental Protection Agency (EPA) and the Internal Revenue Service (IRS) also serve as key contributors in this Federal initiative.

DOT Research and Achievements

In FY 2009, the Department of Transportation conducted hydrogen research, worked to develop the necessary safety codes and standards including emergency response training, and undertook other activities to better understand the impacts of hydrogen-fueled vehicles. The most active of these were the Research & Innovative Technology Administration (RITA), the Pipeline & Hazardous Materials Safety Administration (PHMSA), the National Highway Traffic Safety Administration (NHTSA), and the Federal Transit Administration (FTA).

RITA

RITA was appropriated $0.5M in FY 2009 for hydrogen research focused on creating the Federal and international standards needed to ensure the safe transport of hydrogen and effective response by the police and other emergency workers where this fuel is being used. The bulk of these funds supported safety codes and standards work. They were used to develop training materials for emergency responders and training police and firefighters.

PHMSA

Pipeline Safety Office

PHMSA is the primary Federal regulatory agency responsible for ensuring the safe, reliable and environmentally sound transportation of energy products by pipeline including hydrogen. PHMSA's Pipeline Safety Office has been regulating pure hydrogen gas pipelines since 1970 via 49 CFR Part 192. There are approximately 700 miles of DOT-regulated hydrogen transmission pipeline. Hydrogen pipelines were included as part of the integrity management requirements in 2003 to bolster the awareness of threats to safety and the continuity of service for these lines.

Partnerships between PHMSA's Pipeline Safety Office, pipeline industry operators and partners, other Federal and State agencies and the emergency first responder community are rapidly addressing infrastructure challenges and removing the technical and regulatory barriers for transportation of some alternative fuels. These initiatives are critical for enabling alternative fuel usage to grow nationwide and reach government production targets.

The Office works to ensure that hydrogen is transported safely, even though its hydrogen-related expenditures, particularly for research and development (R&D), are relatively modest compared to other organizations, both government and private. Given that PHMSA is responsible for ensuring pipeline safety, its work is likely to be a key factor in the successful and timely commercialization of hydrogen as an energy carrier. The results of R&D funded by others will provide most of the inputs for establishing the codes and standards used for these regulations.

As hydrogen moves from concept to reality and the public depends on hydrogen availability to meet significant power and/or transportation energy demands, the ability to safely and reliably transport and store larger quantities will become increasingly important. Currently, existing hydrogen pipelines mostly serve industrial demand and hydrogen is transported at constant, relatively low pressure. Confidence in the design, materials of construction, and performance of hydrogen pipelines should remain consistent regardless of the number of miles of pipeline. Given the public's stake in the uninterrupted movement of commodities throughout the nation, the ability of the hydrogen infrastructure to withstand natural disasters and accidents is a major agency interest area.

PHMSA has identified the following nine critical hydrogen research and development gaps in technology, general knowledge or codes and standards that could potentially delay a hydrogen economy:

Understanding of the correlations between pressure, temperature and loss of mechanical properties for pipelines used to transport hydrogen;

- ▶ Loss of mechanical properties due to pressure and temperature interactions can lead to failure. Research and testing are needed to provide more definitive guidance for codes and standards developers;

- ▶ Development of an improved knowledge base and understanding related to the transport of compressed hydrogen at pressures above 2,500 psi;

- ▶ Development of improved knowledge base and understanding related to transport of liquefied hydrogen and effects of hydrogen purity;

- ▶ Investigate and validate the loss of fatigue resistance and impact strength in pipelines;

- ▶ Research on fatigue crack growth;

- ▶ Research and testing to improve understanding of the entire pipeline system using high strength steels to enhance performance in a hydrogen environment;

- ▶ Assessment to understand the effects of hydrogen on pipelines currently in use, such as those now used for transporting oil and natural gas;

- ▶ Research on corrosion control, including coatings, cathodic protection, electrical isolation and interference currents, while developing guidelines and standards for purging, cleaning and maintaining hydrogen pipelines.

In FY2009 PHMSA, Pipeline Safety Office issued a public research solicitation to further address alternative fuel gaps including the nine identified above. All PHMSA research is coordinated with the American Society of Mechanical Engineers (ASME). ASME is crafting a new piping standard addressed in these areas. To facilitate communications about hydrogen pipelines, PHMSA also has created the DOT/PHMSA Pipeline Safety Stakeholder Communication website, which can be found at: |
http://primis.phmsa.dot.gov/comm/hydrogen.htm

NHTSA

NHTSA initiated a safety research program in 2006 to assess fuel system integrity of hydrogen fuel cell vehicles (HFCVs) in crashes. Current Federal motor vehicle safety standards (FMVSS) set performance criteria for fuel system crash integrity for vehicles using liquid fuels, compressed natural gas, and battery drive systems. However, these standards do not currently exist for hydrogen fueled vehicles despite industry interest to facilitate their introduction into the marketplace.

To this end, NHTSA has initiated a research program to assess the safety performance of HFCV fuel systems under similar crash conditions to those prescribed in the existing FMVSS, and to identify and assess any additional life-cycle safety hazards imposed by these unique propulsion systems. Examples of such hazards are rapid release of chemical or mechanical energy due to rupture of high pressure hydrogen storage and delivery systems, fire safety issues, and electrical shock hazards from the high voltage sources, including the fuel cell stack and ultracapacitors. This research supports possible rulemaking to set minimum performance requirements to prevent leakage, fire, or rupture caused by failure of the hydrogen containment system and to prevent electrical shock caused by loss of electrical isolation of the high voltage system.

NHTSA also is working through the auspices of the United Nations on an internationally harmonized safety regulation for hydrogen fuel cell vehicles – UN/ECE WP29.

FTA

FTA is working with DOE on about $14 million in demonstration projects to better understand the performance of fuel cell and hydrogen commercial vehicles in real operating conditions.

DOE Research and Achievements

The U.S. DOE Hydrogen Program works in partnership with industry, academia, national laboratories, Federal and international agencies to:

- Overcome technical barriers through research and development of fuel cell technologies for transportation, distributed stationary power, and portable power applications, as well as hydrogen production, delivery, and storage technologies;

- Address safety concerns and develop model codes and standards;

- Validate and demonstrate fuel cell and hydrogen technologies in real-world conditions;

- Educate key stakeholders whose acceptance of these technologies will determine their success in the marketplace.

DOE's Hydrogen Program is a cooperative effort involving the Offices of Energy Efficiency and Renewable Energy, Fossil Energy, Nuclear Energy, and Science. These offices work with industry, national laboratories, universities, government agencies, and other partners to overcome barriers to the widespread use of fuel cells and hydrogen fuel. Activities include R&D focused on advancing the performance and reducing the cost of these technologies, a market transformation element dedicated to facilitating hydrogen and fuel cell adoption, and activities focused on addressing non-technical challenges such as codes, standards and public awareness. The program addresses infrastructure challenges through its work in developing and improving hydrogen production and delivery methods and via its vehicle and station learning demonstrations.

In the long term, fundamental science is a key component in attacking the technology challenges outlined above. Therefore, DOE funds basic research of relevance to issues underpinning the production, storage and use of hydrogen for advanced energy applications. The topical areas covered are novel materials for hydrogen storage, membranes for separation, purification and ion transport, design of catalysts at the nanoscale, solar hydrogen production, bio-inspired materials and processes, biological hydrogen production and cross-cutting science.

Since 2002, DOE's R&D activities have:

- Significantly reduced the cost of automotive fuel cells (from $275/kW in 2002 to $73/kW in 2008, based on projections of high-volume manufacturing costs);

- Doubled the durability of fuel cell systems in vehicles operating under real-world conditions (data in 2006 showed 950-hour durability—today, this number is more than 1,900 hours, equivalent to approximately 57,000 miles of driving);

- Reduced the cost of producing hydrogen from both renewable resources and natural gas (DOE has validated a projected cost for hydrogen produced at high volume from natural gas of $3.00/gallon gasoline equivalent, which is cost competitive with gasoline when considering the efficiency gains of using a fuel cell);

- Verified compatibility of hydrogen for fiber reinforced polymer pipe for hydrogen pipelines;

- Doubled the capacity of tank trucks for bulk hydrogen delivery (developed manufacturing capability to produce 38 ft by 42 inch diameter cylinders designed for ISO packaging specifications, and have passed burst tests at 3600 psi);

- Successfully opened and operated 20 hydrogen stations (over 88,000 kg of hydrogen produced or dispensed) as part of the Hydrogen Learning Demonstration; and,

- Achieved refueling times of 5 minutes or less for 5 kg of hydrogen at 350 bar.

DOE's R&D roadmap establishes the following technical targets:

▶ For light duty vehicles, reduce the cost of fuel cells to $30/kW;

▶ Develop an automotive fuel cell with 5,000-hour (150,000-mile) durability;

▶ Develop on-board hydrogen storage technologies to enable more than 300-mile driving range across all vehicle platforms, without compromising passenger/cargo space or performance;

▶ For stationary fuel cells, reduce the cost of fuel cells to $750/kW and develop a distributed generation polymer electrolyte membrane fuel cell system with 40,000 hours durability and 40% electrical efficiency; and,

▶ Reduce the delivered cost of hydrogen to $1/gallon gasoline equivalent.

Table 3.1 highlights some of the cutting edge progress the Federal Government and other stakeholders have made.

Table 3.1. Additional Federal & Stakeholder Progress to Date	
Additional Federal & Stakeholder Progress to Date	
Fuel Cell Vehicle Performance and Cost	DOE estimates the cost of automotive fuel cell systems has been reduced by 73%, from $275/kilowatt (kW) in 2002 to $73/kW in 2008. These projections (assuming 500,000 units per year) have been validated by an independent assessment, which concluded that $60 - $80/kW is a "valid estimation" of high-volume manufacturing cost, using 2008 technology. Costs will need to be reduced to about $30/kW to be competitive with gasoline internal combustion engines. Automotive fuel cell durability has also improved, with vehicles in real-world demonstrations showing about 2000-hour durability. Advances in key components (e.g., fuel cell membranes, catalysts, etc.) have enabled laboratory demonstrations of more than 7,300 hours of durability, in single-cell testing. Complete fuel cell "stacks" still need to meet the target of 5000-hour durability under real world conditions, which corresponds to roughly 150,000 miles of driving. General Motors has achieved 3,500-hour durability in dynamometer tests and greater than 5,500-hour durability in the lab. 140 vehicles fuel cell vehicles have been demonstrated through the National Learning Demonstration, traveling over 1.9 million miles. Key results include: fuel cell efficiency of 53 – 58%; durability of nearly 2,000 hours (nearly 60,000 miles); and driving range of 196– 254 miles. Onboard hydrogen storage tanks for vehicles have demonstrated capacities of 2.8-3.8% hydrogen by weight (17-18 grams/liter) at 350 bar and 2.5-4.4% hydrogen by weight (18 to 25 grams/liter) at 700 bar, compared to the 2015 target of 5.5 wt% and the ultimate target of 7.5 wt%. Some manufacturers have reported more than 300-mile ranges with high-pressure tanks, in limited vehicle platforms. Promising materials for low pressure storage have been identified with 50% storage capacity improvement since 2004—advanced materials storage technologies have the potential to enable a 300-mile driving range across all vehicle types.
Vehicle Supply	Through its Technology Validation efforts, DOE has worked with original equipment manufacturers to demonstrate 140 hydrogen fuel cell vehicles in real-world conditions. Additional recent industry demonstrations bring the total number of to more than 200 vehicles demonstrated in the United States. California's ZEV mandate offers a further opportunity to increase deployments of fuel cell vehicles.

Fuel Supply	According to DOE, the cost of producing hydrogen from distributed reforming of natural gas (where hydrogen is produced at the refueling site) has been reduced to $3.00/gge (projected for high-volume production and widespread deployment), reaching the cost target of $2.00 – $3.00/gge. R&D efforts have reduced the cost of other distributed hydrogen production technologies, including electrolysis and reforming of renewable bio-derived liquids (current cost estimates are $4.50 – 5.00/gge) and longer-term renewable pathways (current cost estimates are $5.00 – 9.00/gge for large-scale production at centralized facilities, which includes $3.00/gge for delivery). However, the cost of hydrogen from renewable and other pathways must still be reduced to $2:00- $3:00/gge. Hydrogen delivery costs have also been reduced: projected delivery costs using tube-trailers have been reduced by about 30% since 2005 to ~$4.00/gge; and projected delivery costs using pipelines have been reduced more than 10% to less than $3.50/gge, compared to the target of <$1/gge. Commercially available fiber-reinforced polymer (FRP) pipes show no degradation or leakage from high-pressure hydrogen.
Refueling	Refueling time of 5 minutes or less for 5 kilograms (kg) of hydrogen at 350 bar dispensing pressure has been achieved, meeting DOE's near-term target. More than 90,000 kg of hydrogen have been produced and dispensed, as reported by the National Renewable Energy Laboratory.
Public Perception & Education	DOE and stakeholders have been conducting a broad range of activities to educate key audiences, including safety & code officials, State and local government officials, end-users and early adopters, students, and local communities where demonstration projects will take place. Progress to date includes: launch of a training program for first-responders (a hands-on course and an online course that has had more than 9,000 users); launch of an Internet course for code officials; workshops and seminars held to help decision-makers identify opportunities for fuel cell deployments (this involved partnerships with State and State /regional hydrogen and fuel cell initiatives); launch of the "*Increase Your H2IQ Public Information Program*" (includes radio spots, podcasts, print materials, and a MySpace page); and middle school and high school curricula and teacher professional-development programs, which have reached more than 7,000 teachers since 2004.
Insurance & Liability	DOE and its stakeholders have held workshops to develop approaches to address insurability concerns for suppliers and users. Insurance representatives have met with advisory panels and government representatives to help identify insurance related issues.
Standards, Codes & Permitting	DOE, DOT and stakeholders have developed a national template identifying the key codes necessary for hydrogen and fuel cells and the code-development organizations responsible for them. A comprehensive document of hydrogen codes ("Hydrogen Technologies Code"—NFPA2) has been developed to compile all existing hydrogen-related codes and add new critical codes as they are developed (document is currently under review—due for release in 2010). To date 22 hydrogen codes and standards have been published—28 are under preparation/review, and an international draft standard for fuel quality is expected to be published in the fall of 2009. A number of stakeholders are facilitating the development of codes and standards by supporting research and validation necessary to provide the data needed for technically sound codes and standards. DOE has developed Web-based resources and performed extensive education and outreach to facilitate the permitting process. 15 permitting workshops have been conducted since 2007, training 250 code officials.

Safety & Emergency Response	Basic training for first-responders has been initiated. DOE developed an online course, "Introduction to Hydrogen Safety for First Responders," which has registered more than 9,000 users since its launch in 2007. An advanced course for first-responders has also been launched, incorporating hands-on training.
Sustained Federal Commitment	The Federal Government has been funding R&D programs for developing relevant technologies, fuels and fuel supplies. Federal policies are being developed to support and facilitate the development of business cases to support private sector investment. Tax credits and other financing tools can support vehicle conversion and infrastructure development.
Sustained Marketplace Commitment	In early 2008 General Motors launched Project Driveway, which has resulted in the deployment of 100 fuel cell vehicles for a consumer test market in the United States. A similar Project Driveway program was launched in Europe in November 2008. In July 2008, Honda began leasing fuel cell vehicles to a limited number of retail consumers in Southern California, with plans to deploy 200 vehicles by the summer of 2011. Toyota, Honda, GM, Hyundai, and Daimler have all announced plans to commercialize fuel cell vehicles by 2015. Proterra has announced plans to commercialize a fuel cell bus by 2012.
Source: U.S. Department of Transportation, Research & Innovative Technology Administration and U.S. Department of Energy, Office of Efficiency and Renewable Energy's Fuel Cell Technologies Program.	

USDA Research and Achievements

USDA incorporates its hydrogen activities throughout its alternative fuels programs. USDA's alternative fuels efforts are focused through four programs.

The first is Rural Energy for America Program (REAP) promotes energy efficiency and renewable energy for agricultural producers and rural small businesses through grants and loan guarantees. Eligible renewable energy projects include commercially available wind, solar, biomass and geothermal; and hydrogen derived from biomass or water using wind, solar or geothermal energy sources. Congress has allocated it: $55 million for FY 2009, $60 million for FY 2010, $70 million for FY 2011, and $70 million for FY 2012. USDA is developing regulations to implement the program.

The second USDA program is Agriculture and Food Research Initiative (AFRI) Competitive Grants Program that is funding an effort to develop a system for the biological production of hydrogen from agricultural resources.

The third program is the Agricultural Research Service (ARS). ARS research serves to bring coordination, communication, and empowerment to about 1,000 projects including those affecting hydrogen and climate change. Every single ARS research project is peer reviewed by scientific panels arranged through its Office of Scientific Quality Review.

The final USDA program is the Small Business Innovation Research (SBIR) Program. Its Biofuels and Biobased Products Program Area also is developing technology to produce hydrogen from biomass. Eligibility is limited to small (less that 500 employees) US owned and operated businesses. Its Rural Development Program Area supports research and development of technologies that will provide hydrogen-generated power to rural communities.

EPA Research & Achievements

The EPA's National Vehicle and Fuel Emissions Laboratory in Ann Arbor, Michigan participated in the early Hydrogen Economy efforts by installing hydrogen refueling infrastructure at their facility, and partnering with United Parcel Service and DaimlerChrysler in demonstrating a fuel cell urban delivery truck. This demonstration lasted for three years and successfully demonstrated both vehicle and refueling operations in colder climates. During this period the EPA also developed safe fuel economy testing capability and protocols, and officially certified the first fuel cell vehicle in America, the Honda FCX. The UPS/DaimlerChrysler partnership is now concluded, and EPA is concentrating its hydrogen activities in vehicle certification testing and continued participation in the California Fuel Cell Partnership.

Federal Interagency Coordination

To better leverage Federal research and other activities, a staff-level Hydrogen and Fuel Cell Interagency Working Group (IWG) under the National Science and Technology Council has held monthly meetings since 2003. More than 10 Federal agencies use the IWG as a forum for sharing research results, technical expertise, and lessons learned about hydrogen and fuel cell program implementation and technology deployment. The IWG also facilitates coordinating related projects to ensure efficient use of taxpayer dollars. IWG members include the Departments of Agriculture, Commerce, Defense, Energy, Homeland Security, State and Transportation, Environmental Protection Agency, National Aeronautics and Space Administration, National Science Foundation, Office of Management and Budget, Office of Science and Technology Policy, and U.S. Postal Service.

Energy Policy Act of 2005 (EPACT, section 806) mandated the creation of the Hydrogen and Fuel Cell Interagency Task Force (ITF). The ITF is comprised of senior-level representatives from the agencies participating in the IWG. The ITF has the ability to make departmental decisions that can influence the development and implementation of hydrogen and fuel cell programs. Work resulting from decisions by the ITF is complemented and supported by the staff-level IWG. To date, the ITF has focused its efforts on Federal leadership of early technology adoption and opportunities for interagency partnerships to demonstrate and deploy hydrogen fuel cell technologies in early market applications. In 2005, the task force created a website at www.hydrogen.gov to provide information on all Federal hydrogen and fuel cell activities.

Constructing an Infrastructure Plan

RECENT STUDIES RELATED TO A HYDROGEN INFRASTRUCTURE

Federal agency, industrial and academic stakeholders have and continue to develop frameworks and implementation plans for the transition from a transportation system based on fossil fuel to developing advanced technologies from domestic renewable energy resources that produce the greatest GHG reductions and other environmental benefits. The most pertinent include:

The National Academies' National Research Council – *Transitions to Alternative Transportation Technologies – A Focus on Hydrogen, 2008*

This study estimates the resources needed to bring hydrogen fuel cell vehicles (HFCVs) to the point of competitive self-sustainability in the marketplace. It also projects the impact on oil consumption and carbon dioxide emissions as HFCVs become a large fraction of the light-duty vehicle fleet.

The National Academies' National Research Council – *The National Academies Summit on America's Energy Future: Summary of a Meeting, 2008*

This summit brought together many energy experts to discuss how U.S. energy needs can be met without irreparably damaging Earth's environment or compromising U.S. economic and national security. It is part of the ongoing project "America's Energy Future: Technology Opportunities, Risks, and Tradeoffs," providing authoritative estimates and analysis of the current and future supply of and demand for energy; new and existing technologies to meet those demands; their associated impacts; and their projected costs.

The National Academies' National Research Council – *Sustainable Critical Infrastructure Systems: A Framework for Meeting 21st Century Imperatives, 2009*

This report discusses the essential components of a new paradigm for the renewal of critical infrastructure systems, and outlines a framework to ensure that ongoing activities, knowledge, and technologies can be aligned and leveraged to help meet multiple national objectives.

The National Academies' National Research Council – *Review of DOE's Nuclear Energy Research and Development Program, 2008*

The FY 2006 Budget funded a National Academy of Sciences review of DOE's Nuclear Energy research programs to recommend priorities for those programs given the likelihood of constrained budget levels in the future. The programs to be evaluated were Nuclear Power 2010, the Generation IV reactor development program, the Nuclear Hydrogen Initiative, the Global Nuclear Energy Partnership (GNEP)/Advanced Fuel Cycle Initiative (AFCI), and the Idaho National Laboratory facilities program. The committee's evaluation of each is summarized in this report, along with its assessment of program priorities and oversight and its relevant recommendations.

National Hydrogen Association – *The Energy Evolution: an analysis of alternative vehicles and fuels to 2100, April 2009*

The Energy Evolution compares more than 15 of the most promising fuel and vehicle alternatives over a 100-year period, in scenarios where a mix of vehicles is used initially with one fuel and vehicle alternative becomes dominant in the vehicle mix over time. The scenarios evaluate the performance and viability of each alternative in terms of greenhouse gases, oil imports, urban air pollution and societal costs.

Department of Energy's Oak Ridge National Laboratory – *Analysis of the Transition to Hydrogen Fuel Cell Vehicles and the Potential Hydrogen Energy Infrastructure Requirements,* March 2008

Achieving a successful transition to hydrogen-powered vehicles in the U.S. automotive market will require strong and sustained commitment by hydrogen producers, vehicle manufacturers, transporters and retailers, consumers, and governments. In response to the EPACT of 2005 requirement and recommendations by the National Academies of Science, DOE's Hydrogen, Fuel Cells and Infrastructure Technologies Program (HFCIT) has supported a series of analyses to evaluate alternative scenarios for deployment of millions of hydrogen fueled vehicles and supporting infrastructure. This report shares the results of those analyses.

Department of Energy Hydrogen Program – *Hydrogen, Fuel Cells & Infrastructure Technologies Program, Multi-Year Research, Development and Demonstration Plan, Planned program activities for 2005-2015,* updated April 2009

This Plan details the goals, objectives, technical targets, tasks and schedule for Energy Efficiency and Renewable Energy Program's contribution to the DOE Hydrogen Program – the Hydrogen, Fuel Cells and Infrastructure Technologies Program. Similar detailed plans exist for the other DOE offices and can be found at http://www.hydrogen.energy.gov. The DOE Hydrogen Posture Plan is the integrated plan for all four offices and can be found at:
http://www.hydrogen.energy.gov/pdfs/hydrogen_posture_plan_dec06.pdf.
(http://www1.eere.energy.gov/hydrogenandfuelcells/mypp/)

Department of Energy and Department of Transportation – *Hydrogen Posture Plan: An Integrated Research, Development and Demonstration Plan (2006)"*

The Hydrogen Posture Plan outlines the DOE role in hydrogen energy research and development, in accordance with the former Administration's National Hydrogen Energy Vision and Roadmap. It also lays the foundation for a coordinated response, including collaboration with the DOT, to the former President's plan for accelerating implementation of hydrogen infrastructure and fuel cell technologies. (http://www.hydrogen.energy.gov/pdfs/hydrogen_posture_plan_dec06.pdf)

Department of Transportation – *U.S. Department of Transportation Roadmap for the Safety of Hydrogen Vehicles and Infrastructure to Support a Hydrogen Economy,* October 2005

The DOT Hydrogen Roadmap is helping to guide DOT Hydrogen Safety Research, Development, Demonstration, and Deployment (RDD&D) programs. It outlines the roles and activities of each participating operating administration and their parallel efforts within the DOT. The Roadmap also serves as an outreach document for communication, coordination, and collaboration with other Federal agencies, industry, the public, and Congress.

Department of Transportation – *Hydrogen Infrastructure Safety Technical Assessment and Research Results Gap Analysis,* April 2006

To enable successful introduction of hydrogen into the marketplace, the development of appropriate technical codes, standards, and regulations providing high levels of safety and environmental protection should proceed in parallel with the substantial pace of new technology development. This report identifies gaps in the current hydrogen technology base, and recommends solutions to U.S. DOT for closing these gaps.

Department of Transportation – A Policy Framework for Addressing Risk during Transition to the Hydrogen Economy, (draft), August 2006

This framework provides an analytic basis for PHMSA's Office of Hazardous Materials Safety (OHM) in support of a consistent and effective regulatory response to emerging options for hydrogen fuel delivery and dispensing facilities. The analytic basis provided is intended to serve as a core framework for risk assessment of hydrogen economy transition measures.

Department of Transportation – *Alternative Fuels Roadmap (draft),* January 2009

This report outlines the need to replace petroleum as the source of transportation fuel in the US, the different types of alternative fuels, and what DOT, in conjunction with other organizations, has done and will need to do in the area of alternative fuels.

California Fuel Cell Partnership – *Hydrogen Fuel Cell Vehicle and Station Deployment Plan: A Strategy for Meeting the Challenge Ahead Action Plan,* February, 2009

This action plan details a strategy for deploying hydrogen fueling stations and fuel cell vehicles in California. It specifies the steps needed to meet the fuel needs of 4,300 passenger vehicles and 20 fuel cell buses by 2014, and prepares for even more growth though 2017. The plan calls for 46 retail hydrogen fueling stations in six key California communities at a cost of about $180 million over four years; $60 million from industry and $120 million from government.

National Hydrogen Association – *The Future of Hydrogen: An Alternative Transportation Analysis for the 21st Century,* Webinar, October 23, 2008

This webinar features the analysis and models of 5 to 10 difference alternative fuel and vehicle combinations, recently heard in briefings on Capitol Hill and conferences across the globe. Each combination is analyzed on a 'well-to-wheels' basis and across a wide range of variables, making it one of the most thorough comparisons of next-generation transportation technologies.

General Motors Research & Development Center – *Hydrogen Fueling Infrastructure Assessment,* December 2007

This report demonstrates that a hydrogen fueling infrastructure that could support volume deployment of fuel cell-electric vehicles can be commercially viable and that, in the long term, customers will not have to pay more per mile for hydrogen than they do for gasoline today. Supporting data is provided by key infrastructure stakeholders, including Shell, GE, and DOE.

University of California, Davis – *Optimized Pathways for Regional H2 Infrastructure Transitions: A Case Study for Southern California,* January 2008

Southern California has been proposed as a likely site for developing a hydrogen refueling infrastructure. This paper applies dynamic programming to identify optimized strategies for supplying hydrogen over time in Southern California.

WHERE IS A NATIONAL PLAN?

There is no single national plan for building a hydrogen infrastructure. There are national plans orchestrating each of the numerous activities that constitute a hydrogen infrastructure. Just as authority and responsibility for all of these elements is disbursed across the public and private sectors, coordination efforts are focused on the process owners and stakeholders who make the largest difference in achieving disparate goals. As reflected by the studies cited above, there are national, and sometimes international, efforts addressing each of the key aspects such as safety codes and standards, Federal research and development, requirements for station siting, providing outreach and tools for State and local decisionmakers. The work that is being done today is providing the context that will make a national framework possible.

However, if the nation is to achieve the goal of a greener, more secure economy, there is a need for a long term focus on alternative fuels research, development and deployment including hydrogen. This is not solely a Federal responsibility. State, local and private sector stakeholders will be key to leveraging and realizing a common commitment for this fundamental change in American mobility. Accomplishing this transition will be no less impressive than building a transcontinental railroad or the Interstate highway system.

There is a strong foundation of work to start this journey to widespread use of hydrogen-fueled vehicles. Key publications include:

▶ A listing of about 60 hydrogen refueling stations across the nation, which is included as Appendix C;

▶ The California Fuel Cell Partnership action plan
(http://www.fuelcellpartnership.org/sites/files/Action%20Plan%20FINAL.pdf);

▶ California's California Hydrogen Highway
(http://www.hydrogenhighway.ca.gov/blueprint/blueprint.htm).

In addition, DOE's Oak Ridge National Laboratories has assessed how best to create a hydrogen infrastructure and identifies implementation strategies including market supports lasting until 2025 to make the technology more attractive to consumers http://cta.ornl.gov/cta/Publications/Reports/ORNL_TM_2008_30.pdf
The lab has conducted scenario analyses to predict market behaviors

http://www.hydrogen.energy.gov/pdfs/progress08/x_1_greene.pdf

The NRC work on transitioning to hydrogen http://www.nap.edu/catalog.php?record_id=12222 and the nation's energy future http://www.nap.edu/catalog.php?record_id=12450 also bring a good grounding for future planning.

The Future

In the past few years, DOT, DOE, and their industrial and academic partners have made significant in advancing the technologies on the path to validation and eventually commercialization. Notable improvements were reported and independently verified for the performance and costs of fuel cells, the capacity of on-board hydrogen storage, and hydrogen fueling technology.

Supported by these frameworks and implementation plans, senior decisionmakers face choices as they reconcile and integrate these and other accomplishments into a path forward for an alternatively fueled America. One important task will be to prioritize and, in some cases, harmonize all of the short, medium and long-term milestones this transition involves and then achieve them. However, this is not solely a Federal responsibility. State, local and private sector stakeholders are key to realizing this fundamental change in American mobility. Accomplishing this transition will be no less impressive than building a transcontinental railroad or the Interstate highway system.

These choices include:

▶ **Technology**

- **Innovations** to increase the supply, efficiency, range and cost competitiveness of fuel cell vehicles, and reduce the cost of producing hydrogen from domestic resources using green production methods.

▶ **Public & private sector organizational**

- *Land use and station siting* guidance to ensure the safe and efficient development of this new infrastructure including development of future improvements to reduce the size of the current station footprint.

- *Public education and outreach* to increase awareness, motivate key stakeholders, and facilitate the acceptance of the new technology.

▶ **Commercial sector**

- *Market development and deployment* including policy decisions about whether implementation should focus on growing urban and regional markets where there is likely to be strong consumer demand or on a national network so that vehicles can operate regardless of location.

- *Partnerships* to bring together the stakeholders whose collaboration is essential to the deployment of hydrogen vehicles and a hydrogen infrastructure, i.e., Federal , State , and local government, automakers, fuel providers, electricity producers, other relevant industries, academia, environmental groups, and the public.

▶ **Safety codes & standards**

- **Universally accepted requirements** to establish the appropriate safety, quality and consumer protection also be provided to match fossil fuel standards including the safety of compressed hydrogen (CH2) and liquid hydrogen (LH2) fueled vehicles and subsystems, of fueling infrastructure and of fueling interfaces, as well as safe integration and compatibility with mixed fleet and fuels operations during a long transition period.

- *Emergency response training* to provide the knowledge and tools first responders will need to deal with the different dangers hydrogen presents as well as provide the regulatory requirements needed to address the new technologies and innovations this transition will generate.

▶ **Sustained commitment**

- ***Programs and incentives*** to address the expected cost differentials between hydrogen vehicles and conventional vehicles during the transition period. Some of these activities should be coordinated with the safety, codes and standards activities in order to accelerate the insurance industry's adoption of comparable rate structures and procedures.

Appendix A—Infrastructure Maps

Map 1: LNG Facilities in the United States as of June 2004

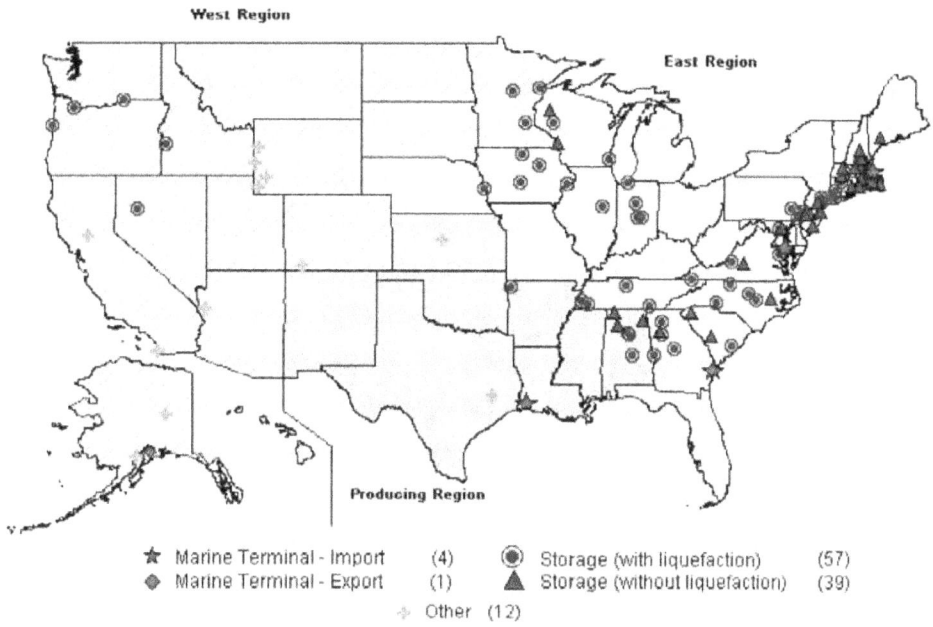

Source: U.S. Department of Energy, Energy Information Administration, U.S. LNG Markets and Uses: June 2004 Update, Page 2, http://www.eia.doe.gov/pub/oil_gas/natural_gas/feature_articles/2004/lng/lng2004.pdf

Map 2: U.S. Gas Transmission Pipelines

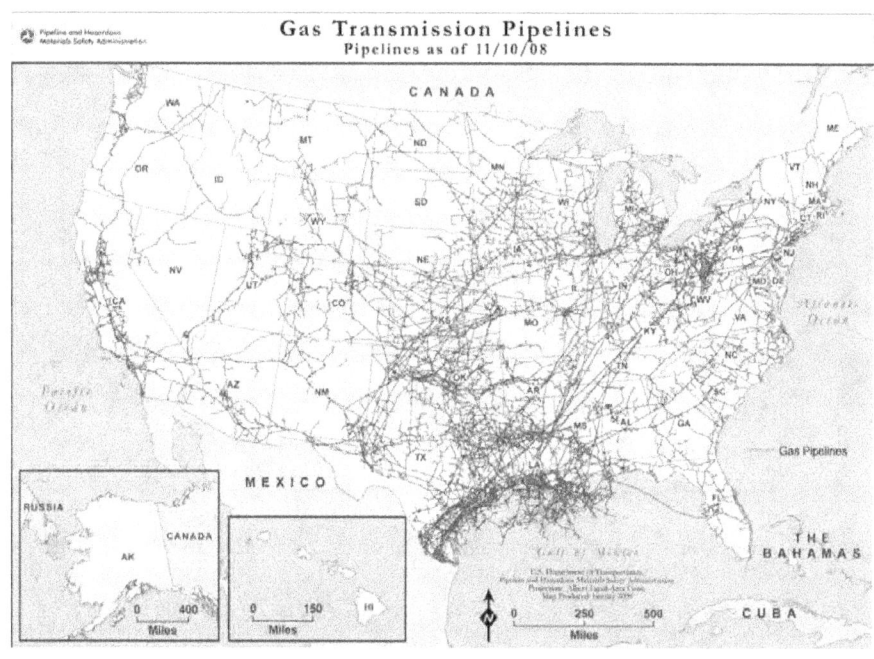

Source: U.S. Department of Transportation, Research and Innovative Technology Administration and Pipeline and Hazardous Materials Safety Administration, *Alternative Fuels Roadmap*, January 2009, p. 2-64

Map 4: Geographic Distribution of Natural Gas Processing Plants

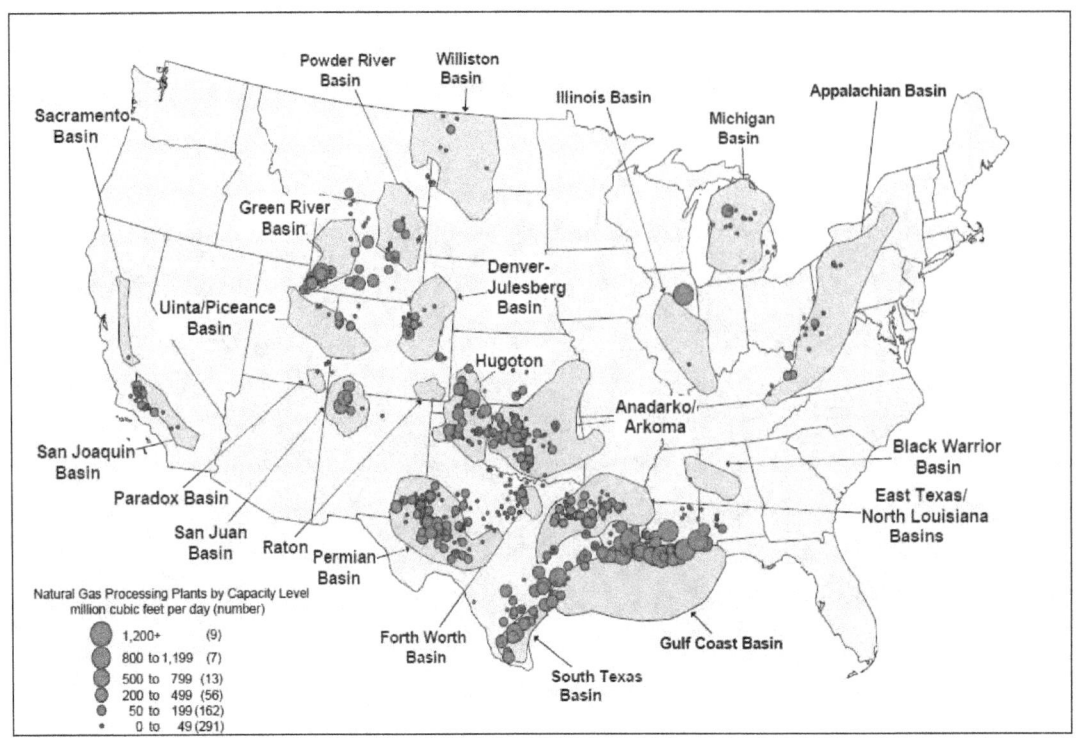

Source: U.S., DOE, "Natural Gas Processing: The Crucial Link Between Natural Gas Production and Its Transportation to Market," January, 2006, http://www.eia.doe.gov/pub/oil_gas/natural_gas/feature_articles/2006/ngprocess/ngprocess.pdf.

Map 5: Geographic Distribution of Electric Charging Stations in the United States by State

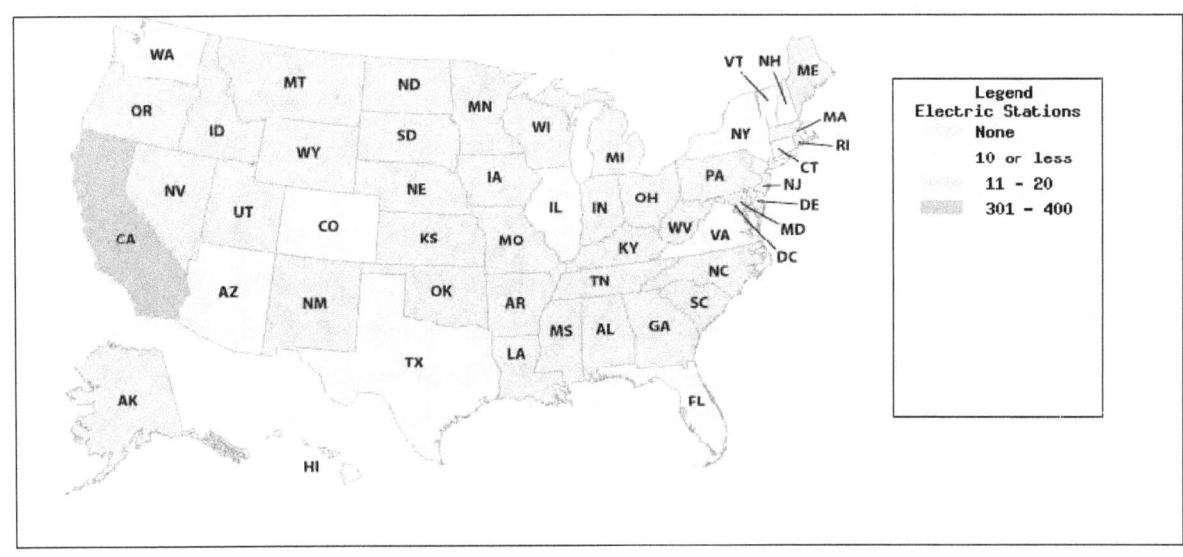

Source: U.S. Department of Transportation, Research and Innovative Technology Administration and Pipeline and Hazardous Materials Safety Administration, *Alternative Fuels Roadmap*, January 2009, p. 2-74.

Map 6. Geographic Distribution of Propane Service Stations in the United States by State

Source: Department of Energy.

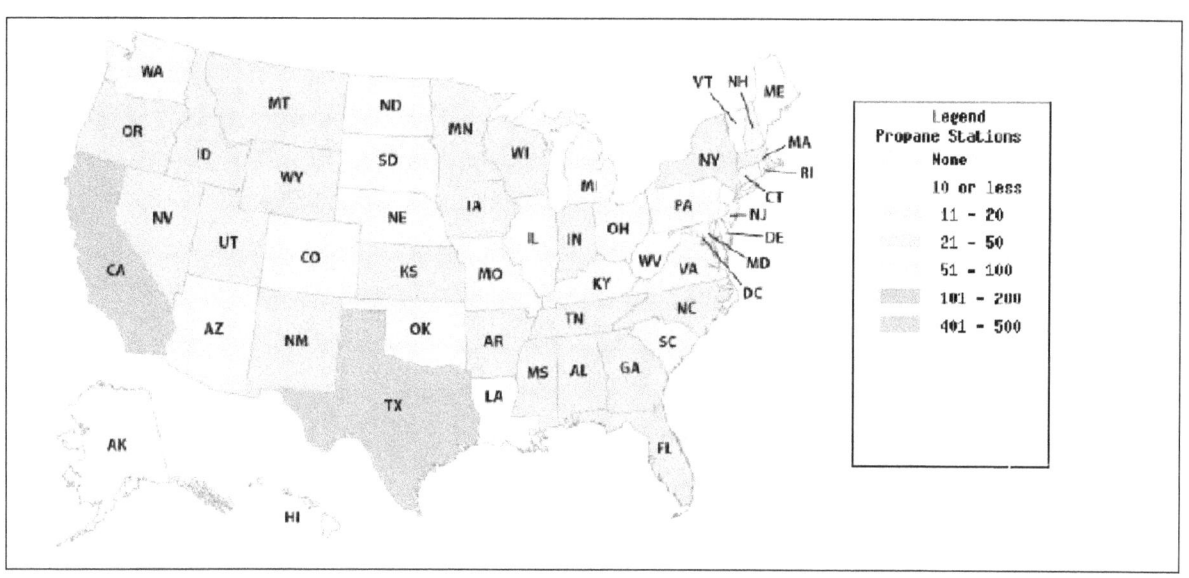

(Prepared by RITA's Volpe Center)

Appendix B—Comparison of Alternative Fuels Technical Readiness

Fuel	Production	Vehicular Demand	Capacity	Major Impediments to Future Use
Gasoline	Produced using domestic and foreign crude oil. Considerable quantities must be imported to meet domestic demand.	Enormous and slowly growing.	The United States has sufficient proven reserves of crude oil to last at least 12 years. The United States has insufficient refinery capacity to produce all of the gasoline that it consumes.	Declining reserves of crude oil Production of greenhouse gas and other pollutants Use of foreign sources of supply of crude oil and gasoline Price volatility of crude oil and gasoline
Diesel	Produced using domestic and foreign crude oil.	Enormous and slowly growing.	See Gasoline	See Gasoline
Electricity	Produced with both nonrenewable fuels (e.g., oil and coal) and renewable sources of energy (e.g., solar, wind, and water power).	Pure electric (including plug-in hybrids): relatively small and static. Other: gasoline / battery hybrids currently in high demand.	Renewable sources currently in use have limited capacity.	Limited battery capacity Limited fueling infrastructure Limited availability of electric vehicles Speed of implementing smart grid technologies Primary sources of lithium for batteries are foreign
Natural Gas	Despite vast reserves of gas, some importation, particularly from Canada, occurs. During the heating season, some imports from overseas are brought in by ship to augment domestic supplies.	Large and growing.	Based on a comparison of proven reserves with gross withdrawals, the United States has sufficient proven reserves to last over 90 years. If demand were to increase due to increased use as a motor fuel, additional processing-plant capacity might be needed.	Limited fueling infrastructure Limited availability of natural gas powered vehicles

Fuel	Production	Vehicular Demand	Capacity	Major Impediments to Future Use
LPG	Produced about 50/50 from natural gas processing and petroleum refining. Production a function in part of the demand for natural gas and other refining products.	Large but declining.		Limited fueling infrastructure Limited availability of natural gas powered vehicles
LPG	Produced about 50/50 from natural gas processing and petroleum refining. Production a function in part of the demand for natural gas and other refining products.	Large but declining.		Limited fueling infrastructure Limited availability of natural gas powered vehicles
Ethanol	In 2007, 6.5 billion gallons produced at 134 plants in 26 States. In Sept. 2008, capacity was 9.2 billion gallons per year. Domestic production growing.	E85: Growing. Ethanol in gasohol: very large and rapidly growing.	In 2007, 77 new or expanded plants being planned. In 2007, 7.8 billion gallons and slated to increase in the near term. As of late 2008, many plants are closing or being postponed.	Fuel – food trade-off Limited fuel availability due to numerous distribution challenges. Engine compatibility concerns Various concerns relating to using the existing petroleum pipeline system for distribution Uncertainties associated with the development of cellulosic ethanol production capabilities
Biodiesel	As of late 2008, produced at 176 plants in 40 States. Production tripled between 2004 and 2005 and tripled again between 2005 and 2006. In 2008, a number of plants idle to reassess profitability.	Large and very rapidly growing.	Capacity is currently at about 2.6 billion gallons per year. Adequate feedstocks exist to support production of 1.7 billion gallons per year.	Diesel engine compatibility concerns Limited fuel availability due to numerous distribution challenges. Rethinking feedstock choice due to price increases. Fuel – food trade-off
Hydrogen	Serious production for fuel still in early stages. Debate over relative costs and benefits associated with centralized and decentralized production as yet unresolved.	Minor but growing.		Limited fueling infrastructure; limited pipeline distribution infrastructure; limited carrying capacity of tanks on tank trucks and the need for Federal regulatory approval to increase that capacity; limited availability of hydrogen powered vehicles.

Appendix C—International Hydrogen and Fuel Cell Infrastructure Development for Transportation Applications

(Note: currency converted using April 20, 2009 conversion rates)

Hydrogen and fuel cell infrastructure development is a priority around the world. Below is a summary of activities in key countries:

Canada

- Canada currently has 16 hydrogen filling stations that are used by 20 fuel cell buses, 10 hydrogen ICE shuttle buses, 4 hydrogen/compressed natural gas transit buses, 5 fuel cell light-duty vehicles, and 12 ICE pick-up trucks.

- British Columbia is working with through a public-private partnership to develop the British Columbia Hydrogen Highway. It will accelerate the commercialization of hydrogen and fuel cell technology and creating a legacy of economic growth and environmental benefits through three phases of implementation:

    - Phase One (2004-2007) built 6 stations and purchased 20 fuel cell buses for use in Whistler for the 2010 Winter Olympics.

    - Phase Two (2008-2009) is focusing on the delivery and operation the Phase One infrastructure and educating the public on technology.

    - Phase Three (2010 and beyond) will focus on expanding the infrastructure to 30 – 50 stations in order to service large-scale deployments of vehicles in the 2015 timeframe.

Japan

- Japan has a significant Hydrogen and Fuel Cell Demonstration Project that includes development of vehicle infrastructure. The project is coordinated by the Ministry of Economy, Trade and Industry (METI). Phase 1 of the fuel cell vehicle demonstration project was from 2002 – 2005 and Phase 2 started in 2006 and will run through 2010. In 2008:

    - The project included 43 fuel cell vehicles, 5 fuel cell buses and 12 hydrogen internal combustion engines (ICE) vehicles using 11 fueling stations.

    - There are 9 vehicle manufacturers and 16 hydrogen infrastructure suppliers involved in the project.

    - Over 3,300 stationary residential Proton Exchange Membrane fuel cells are operating as part of the stationary fuel cell demonstration project.

- Currently forming a Low-Carbon Mobility Committee that will develop and steer social demonstration plans in the projected deployment of hydrogen and fuel cell vehicle infrastructure from 2011 to 2015, and will study deployment measures, systems and legislation to be put in place for commercialization to begin in 2015.

- Current plan is for hydrogen infrastructure to be built prior to vehicle introduction.

    - By 2020, 1000 hydrogen fueling stations will serve the anticipated 50,000 fuel cell vehicles manufactured per year.

    - In 2030, 5000 stations will support 1,000,000 fuel vehicles per year production.

European Commission

- The European Commission launched the European Hydrogen and Fuel Cell Joint Technology Initiative (JTI) in October 2008. The JTI is a public-private partnership designed to facilitate and accelerate the development and deployment of cost-competitive, world class European hydrogen and fuel cell based energy systems and component technologies for applications in transport, stationary and portable power.
    - ° The JTI budget from 2008 – 2013 is 470 Million Euro ($613.3 Million) of government funding that requires a 100% industry matching.
- In September 2008, the European Parliament passed a regulation that implements a simplified process for hydrogen vehicle approvals. The objectives of the regulation are to unify requirements in all 27 States for hydrogen fueled vehicles, treat hydrogen vehicles the same as conventional vehicles and to ensure the same level of safety as conventional vehicles.

Germany

- The German National Hydrogen and Fuel Cell Technology Program (NOW) is a joint program funded by four German Federal Ministries. The NOW funds research, development and demonstration activities, including hydrogen infrastructure construction, in order to facilitate market penetration of hydrogen and fuel cell technologies. The government budget for NOW from 2007 – 2016 is 700 Million Euro ($913.4 Million), with a cost share from industry of an additional 700 Million Euro ($913.4 Million).
- Germany's hydrogen and fuel cell technology demonstration and infrastructure activities currently include 7 hydrogen filling stations, 20 hydrogen buses, 15 fuel cell vehicles, and 2 hydrogen internal combustion engine vehicles.
    - ° These demonstration and infrastructure projects were funded by the Germany federal government and the state of North Rhine Westfalia (NRW) from 2003 - 2008.
    - ° The NRW infrastructure is part of the on-going NRW Hydrogen Hyway program, which consists of 40 sub-projects in 9 locations around the state of NRW and one location in Belgium. The planned activities encompass the whole range of hydrogen utilization, including transport, stationary, and special market applications. The overall budget for these projects is 200 Million Euro ($261 Million) from 2009 to 2011, with the Government of NRW committing approximately 70 Million Euro ($91.3 Million) and the federal government and industry contributing the remaining funds.
    - ° 150 miles of industrial hydrogen pipeline currently operates in NRW. The state government is currently finishing a study of local industries to determine the amount of hydrogen that could be supplied from industrial waste streams for use in hydrogen fuel cell vehicles. This is expected as one means for providing hydrogen fuel for vehicles in the near-term. The study is schedule for completion in June 2009.
- 15 Million Euro ($19.57 Million) of the recent German economic stimulus package is dedicated to increase hydrogen fueling infrastructure. The funding language states that Germany expects to have hydrogen fuel cell vehicles commercialized by 2015 and that hydrogen infrastructure needs to be in over supply at this time. The funding is directed toward the construction of 25 additional hydrogen filling stations in synergy with the expected locations for hydrogen fuel cell vehicle usage. There is expected to be cost share from industry partners but the amount has yet to be determined.
- Zemships (zero emission ships) will be cruising Alster Lake near Hamburg, Germany, this summer. The $6.7 million Zemships project, which is a hydrogen fuel cell ferryboat that will carry 100-passengers across the lake. http://www.hydrogencarsnow.com/blog2/index.php/hydrogen-vehicles/hydrogen-fuel-cell-ferryboats-planned-in-near-future/. Instead of conventional mechanical pistons, the ionic compression is used to compress H2 up to 6,500 psi. "Major advantages of this compressor design are excellent and highly energy-efficient delivery rates, no contamination of the hydrogen gas (very important for fuel-cell applications), less moving parts and a reduced n

Denmark

- As reported by the Hydrogen Link Denmark Association, a recent government climate plan calls for strong investments in hydrogen fuelling stations, allowing for all new car sales in 2025 to be electric and hydrogen only.
  - Further public funding for energy R,D&D are to be doubled to 134 Million Euro ($178 Million) annually, where one third of the funds in the past have been spent on hydrogen and fuel cells.

Scandinavia

- The Scandinavian Hydrogen Highway Partnership (SHHP) constitutes a transnational networking platform that catalyses and coordinates collaboration between three national networking bodies – HyNor (Norway), Hydrogen Link (Denmark) and Hydrogen Sweden (Sweden). The collaboration consists of regional clusters involving major and small industries, research institutions and local/regional authorities.

- Today four hydrogen refueling stations and around 20 vehicles are in operation in Scandinavia with ongoing activities to ensure a further 9 stations in the coming year together with up to 50 vehicles.

- The 2015 goal is to build 15 large-scale production and fueling facilities and 30 satellite stations for smaller volumes and to distribute hydrogen in rural areas. The SHHP intends to have these stations used by 100 buses, 500 cars and 500 specialty vehicles.

International Partnership for the Hydrogen Economy

(IPHE) was established in 2003 as an international institution to accelerate the transition to a hydrogen economy. Each of the following IPHE partner countries has committed to accelerate the development of hydrogen and fuel cell technologies to improve the security of their energy supply, environment, and economy:

Australia	Germany	New Zealand
Brazil	Iceland	Norway
Canada	India	Russian Federation
China	Italy	United Kingdom
European Commission	Japan	United States
France	Republic of Korea	

The IPHE provides a mechanism for partners to organize, coordinate and implement effective, efficient, and focused international research, development, demonstration and commercial utilization activities related to hydrogen and fuel cell technologies. The IPHE provides a forum for advancing policies, and uniform codes and standards that can accelerate the cost-effective transition to a hydrogen economy. It also educates and informs stakeholders and the general public on the benefits of, and challenges to, establishing a hydrogen economy.

Appendix D—List of Nation's Hydrogen Stations

Hydrogen Analysis Resource Center: **Operational Hydrogen Fueling Stations** (Updated January 2009)

State	City or Location	Name of station	Source of Information (see citations at the bottom of the list)	Year Station Opened	Description (see list of acronyms and abbreviations at the bottom of this table)
AZ	Phoenix	Arizona Public Service Alternative Fuel Pilot Plant	1,2,3	2001	Fuels: CGH2 and CNG/CGH2 blends (Hythane- HCNG) with maximum inlet pressure rating of 5,000 psi, CNG with maximum inlet pressure rating of 5,000 psi. Supply via two methods: water electrolysis, and off-site production. Vehicles Served: FCV.
CA	Arcata	Humboldt State University	2,4	2008	Fuel:CGH2;on-site hydrogen generation via electrolysis. System also includes compressor, storage tank, and dispenser. Fuels Toyota Prius converted to run on hydrogen. Vehicle H2 storage at 5000 psig.
CA	Burbank	Burbank Station	1,2,3,4	2006	Fuel: CGH2; on-site hydrogen generation via electrolysis (Proton Hogen 200 electrolyzer). Vehicles Served: Fuel cell vehicles and hydrogen powered internal combustion engine vehicles. Organizations Involved: City of Burbank, SCAQMD, Quantum, Proton Energy Systems, Air Products.
CA	Chino	Chino Station	1,2,3,4,5	2005	Fuel: CGH2; autothermal reformer for generation of hydrogen. Organizations Involved: Hyundai, ChevronTexaco, United Technologies, City of Chino, US Department of Energy, AC Transit.
CA	Chula Vista	Chula Vista Mobile Station	1,2,3,4	2003	Fuel: CGH2 @ 3,600 and 5,000 psi; mobile electrolyzer HySTAT-A 36 Nm³/h (3kg/h) can fuel up to 20-30 cars per day. Vehicles Served: Honda FCX FCV. Organizations Involved: City of Chula Vista, Ford, Toyota, Sunline Transit Agency, Hydrogenics.
CA	Davis	University of California-Davis Fueling Station	1,2,3,4	2004	Fuels: CGH2 @ 5,000 psi to Toyota FC hybrid vehicle, Hythane @ 3,200 to 3,600 psi to buses, Liquid hydrogen produced by Air Products at a reformer plant in Sacramento. Using tanker trailers, LH2 is delivered to a cryogenic storage tank near the refueling station. Organizations Involved: Federal Transit Authority, Toyota, Caltrans, Yolo-Solano Air Quality Management District, Air Products, National Hydrogen Association.
CA	Diamond Bar	AQMD Hydrogen Highway Network Fueling Station (H2 Station.org); Diamond Bar - Pinnacle SCAQMD	1,2,3,4	2004	Fuel: CGH2; produced via electrolyzer. Vehicles Served: Fuel cell vehicles (DaimlerChrysler, Toyota, Honda). Organizations Involved: South Coast Air Quality Management District; Hydrogenics; DaimlerChrysler; Toyota Motor Sales USA, Inc., American Honda Motor.
CA	Irvine	National Fuel Cell Research Center	1,2,3,4	2003	CGH2 is delivered. 2006 upgrades include dispensing at 350 bar (5000 psig), and in 2007 upgrades made for dispensing at 700 bar. Organizations Involved: APCI (Air Products), SCAQMD.
CA	Irvine	National Fuel Cell Research Center	1,4	2005	On-site storage of 3 kg compressed H2. Can fuel 1 to 2 cars per day. Organizations Involved: Hydrogenics; University of California, Irvine.
CA	Long Beach	Long Beach Station	1,4	2007	Fuel: CGH2; supplied by HF 150 mobile refueler at 150 kg storage. Organizations Involved: Air Products.
CA	Los Angeles	Honda Fleet Program (Mobile Refueler)	1,2,3,4	2002	Fuel: CGH2, supplied by mobile trailer. Organizations Involved: Honda, Air Products.

CA	Los Angeles	BP LAX Airport Hydrogen Station	1,2,3,4	2004	Fuel: CGH2; hydrogen is generated on-site by electrolysis. Components: Electrolyzer (Stuart/Hydrogen Systems H2 IGEN 15); Compressor (Praxair, from 100 psig to 6,500 psig); Storage tanks (ASME, 4-18 ft³/vessel water volume; the tanks are arranged to provide cascade fueling); Dispenser (5,000 psi hose and nozzle at 31.1 lpm); Supplemental hydrogen storage via 2,400 psi tube trailer. Organizations Involved: Praxair, BP, SCAQMD, CARB, DOE, LAWA, Hydrogenics.
CA	Los Angeles	Clean Energy LAX Airport Hydrogen Station	8	2008	Fuel: CGH2 and CNG. Site is located at an existing CNG station operated by Clean Energy. GM is a partner on this project and the station will primarily serve a fleet of hydrogen-fueld Equinoxes leased by Virgin Atlantic Airlines.
CA	Oakland	AC Transit ChevronTexaco Hydrogen Energy Station	1,2,3,4,5	2005	Fuel: CGH2; H2 derived on-site from natural gas via steam reforming (ChevronTexaco technology). Vehicles Served: Three 40' fuel cell buses, 10 Hyundai hydrogen FC cars. Organizations Involved: AC Transit (Alameda-Contra Costa Transit District), ChevronTexaco Corp., Chevron Energy Solutions, City of Oakland, Hyundai, Hydrogenics, Quest Air.
CA	Ontario	SCAQMD (Mobile Refueler)	1,2,3,4	2006	Compressed hydrogen delivery; mobile refueler. Vehicles served: Hydrogen vehicles powered by internal combustion engine or fuel cells. Organizations Involved: Air Products, City of Ontario, South Coast Air Quality Management District.
CA	Oxnard	BMW LH2 Refueling Station	1,3,4	2001	Liquid H2 is delivered by Air Products. Organizations Involved: Linde, BMW, Air Products, BP, KASERV.
CA	Riverside	SCAQMD	1,2,3,4	2006	Fuel: CGH2 from electrolysis (Proton Hogen 200 electrolyzers). Vehicles Served: Fuel cell vehicles and hydrogen powered internal combustion engine vehicles -- five Toyota Hydrogen Prius (built by Quantum). Organizations Involved: City of Riverside, SCAQMD, Quantum, Proton E.
CA	Rosemead	Southern California Edison	1,2,4,5	2007	This is the first of six stations that Chevron intends to open and operate under DOE contract. Vehicles Served: small KIA and Hyundai H2 FC vehicles.
CA	Sacramento	SMUD Mobile Fueler	1,3,4	2007	Supports two Daimler-Chrysler vehicles (permanent station targeted for 2007)
CA	San Jose	BAAQMD, Santa Clara Valley Transportation Authority	1,2,4	2004	Fuels: CGH2, 75 kg/h at 35 MPa; CH2 from liquid H2 for buses only; liquid hydrogen delivery (Air Products). Organizations Involved: BAAQMD, Gillig, Air Products, Santa Clara Valley Transportation Authority, San Mateo Transportation Authority, California Energy Commission.
CA	Santa Ana	Santa Ana Mobile Station SCAQMD (Mobile Refueler)	1,3,4	2006	Compressed hydrogen delivery, CH2 from steam methane reformation of natural gas; mobile fueler. Vehicles Served: Fuel cell vehicles and hydrogen powered internal combustion engine vehicles, five Toyota Hydrogen-Priuses (built by Quantum). Organizations Involved: Air Products and Chemicals, City of Santa Ana, SCAQMD.
CA	Santa Monica	Santa Monica Hydrogen Station (part of the SCAQMD program)	1,2,3,4	2006	Compressed hydrogen delivery, CH2 from steam methane reformation of natural gas; mobile fueler. Vehicles Served: Fuel cell vehicles and hydrogen powered internal combustion engine vehicles, 5 Toyota Hydrogen-Priuses (built by Quantum); Organizations Involved: Air Products and Chemicals, City of Santa Monica, SCAQMD, Proton Energy Systems, DOE, Quantum Technologies.
CA	Thousand Palms	SunLine Transit Fueling Station	1,2,3,4	2000	Fuels: CGH2, Hythane; H2 generation with HyRadix Adeo reformer (100 Nm³/h). Vehicles served: Two XCELLSIS ZE-buses, One hydrogen-powered Shelby Cobra. Organizations Involved: Schatz Energy Research Center, Sunline Transit, Hydrogenics.
CA	Torrance	Honda Solar Hydrogen Refueling Station	1,2,3,4	2001	Fuel: CGH2; the hydrogen is produced from solar electricity (PV) and from grid-electricity via electrolysis. Vehicles Served: Honda FCx. Organizations Involved: Honda R&D Company, Ltd.

State	City	Station	Codes	Year	Description
CA	Torrance	Honda Home Energy Station	1,3,4	2003	Fuel: Hydrogen produced via natural gas reformation. Storage: 400 liters @ 420 atmospheres. Vehicles Served: 1 car/day.
CA	Torrance	Torrance Toyota Stuart Energy Station	1,2,3,4	2002	Fuel: CGH2; on-site hydrogen production via electrolysis. Vehicles Served: Toyota's fuel cell hybrid vehicle. Organizations Involved: Stuart Energy (now: Hydrogenics), Toyota, Air Products.
CA	West Los Angeles	Shell Station	2,3,4	2008	Organizations Involved: Shell Hydrogen, DOE, GM.
CA	West Sacramento	California Fuel Cell Partnership (CaFCP) Headquarters	1,3,4	2000	Fuels: LH2, CGH2; Air Products and Praxair deliver liquid H2 where it is stored in liquid H2 tanks. Two gaseous dispensers, one at 3,600 psig the other at 5,000 psig, and one liquid hydrogen dispenser. Hydrogen storage: CGH2 at 6,250 psig, LH2: 17,000 liters (4500 gallons) at 50 psig. Vehicles Served: Fuel Cell Vehicle (Daimler Chrysler, Ford, GM, Honda, Hyundai, Nissan, Toyota, Volkswagen). Organizations Involved: BP, Exxon Mobil, Shell Hydrogen, Chevron Texaco, Air Products, Praxair, California Fuel Cell Partnership.
CT	South Windsor	UTC Power South Windsor Campus	1,2,3	2007	Fuel: CGH2 @ 5,000 psig. Designed to support 2 bus fleet between South Windsor and Harford communities and university. Organizations Involved: UTC Power, CTTransit, Greater Hartford Transportation District.
CT	Wallingford	Proton Energy Systems Fueling Station	1,2	2006	Station is used by CTTransit for hydrogen bus demonstration in the Hartford area. Organizations Involved: Distributed Energy Systems (formally Proton Energy), Air Products.
DC	Washington DC	Shell Benning Road Multi-Fuel Refueling Station	1,2,3	2004	Fuels: CGH2 @ 35 MPa, CGH2 @ 70 MPa (future capability), and liquid H2. Shell generates and distributes hydrogen, trucked-in liquid H2. Vehicles Served: Six GM/Opel FCV vans. Organizations Involved: Shell, GM.
DE	Newark	Air Liquide Newark Station	3,6	2007	Organizations Involved: Air Liquide, US Department of Transportation, University of Delaware.
FL	Orlando	Progress Energy, Chevron/Texaco Orlando International Airport	1,2,3,5	2006	Uses H2Ge system to convert natural gas to hydrogen. Designed to support a fleet of two baggage carriers. Fuels Ford V-10, E450 hydrogen-powered shuttle buses at the Progress Energy Site near Orlando International Airport.
FL	Oveido	Progress Energy-BP hydrogen station	1,2,3	2006	Station is located in Oveido at Progress Energy's Jamestown Operations. Supports fleet of five FC vehicles driven by Progress Energy counselors.
HI	Hickam AFB hydrogen station	Air Force Base FC Bus Demonstration	1,2,3	2006	Hydrogen supplied from a tube trailer. At a later date, a hydrogen generation station will be built. Vehicles Served: a 30-foot flight-crew shuttle bus (FC). Organizations Involved: US Air Force, Hawaiian local government.
HI	Hickam AFB mobile station	HCATT, Hydrogenics, Stuart Energy	1,2,3	2004	Three primary Packaged Operating Modules (PODs), which are modular, deployable hydrogen production and fueling stations, designed and developed by HydraFLX Systems. PODs are crush-proof carbon steel packages for military or commercial transport.
IL	Des Plaines	GTI Hydrogen Fueling Facility	1,2,3	2007	The H2 station can produce H2 from natural gas, ethanol, or electrolysis. Organizations Involved: Gas Technology Institute, DOE.
IN	Crane	Naval Surface Warfare Center	1,2,3	2004	Hydrogenics electrolyzer capable of producing 20 kg/day. H2 used for forklifts.
MA	Billerica	Nuvera Fuel Cells	1,2,3	2008	Fuel: CGH2. Hydrogen made from natural gas on-site. Dispensed at 5000 psig. Daily capacity is 40 kg.
MI	Ann Arbor	EPA, National Vehicle and Fuel Emissions Lab.	1,2,3	2004	Fuel: CGH2. Storage: conventional pressurized cryogenic tank (1,500 gal). Vehicles Served: DaimlerChrysler A-Class F-Cell, Dodge Sprinter F-Cell. Organizations Involved: EPA, Air Products, DaimlerChrysler, United Parcel Service (UPS).
MI	Dearborn	Ford Vehicle Refueling Dearborn	1,2,3	1999	Fuels: liquid H2, CGH2 @ 3,600 and 5,000 psi; supplied by Air Products. Vehicles Served: Ford P2000 fuel cell vehicle.

State	City	Station	Type	Year	Description
MI	Detroit	NextEnergy Center Hydrogen Station	1,2,3	2006	Located in NextEnergy's Microgrid power pavilion. Initially, fuels a Daimler-Chrysler A Class FCV used at Wayne State University. Expected to allow for fueling of Daimler Chrysler FC vehicles.
MI	Milford	DOE VDP General Motors Proving Grounds Lead	1,2,3	2004	Fuel: CGH2 350 bar compressed H2 (5000 psig), upgraded to 700 bar in Feb 2005. Liquid fueling dispenser equipment ordered. Vehicles Served: Concept cars. Organizations Involved: General Motors, Air Products, DOE.
MI	Selfridge	Selfridge Air National Guard Station	1,2,3,5	2007	On-site hydrogen produced via steam methane reforming. The station can deliver up to 80 kilograms per day of gaseous hydrogen from the dispenser at a pressure of 5000 pounds per square inch. This is enough fuel for up to 20 fuel cell vehicles per day. Organizations involved: Chevron Hydrogen, U.S. Department of Energy, the U.S. Department of Defense, Hyundai Kia Motor Company, NextEnergy, and United Technologies.
MI	Southfield	DTE Energy Hydrogen Technology Park	1,2,3	2004	Fuel: CGH2 @ 6,000 psi. On-site generation of hydrogen via electrolysis ($30 Nm^3$ = 65kg/day). Vehicles Served: A-Class "F-Cells" and Sprinter delivery van Generation 1 vehicles. Organizations Involved: DOE, BP, DaimlerChrysler, DTE Energy, State of Michigan, City of Southfield.
MI	Taylor	City of Taylor Hydrogen Station	1,2,3	2006	Fuel: CGH2. Organizations Involved: DOE, Ford, BP. Vehicles Served: 26 fuel cell vehicles (Ford Focus Fuel Cell Generation 1 vehicles).
MO	Rolla	University of Missouri-Rolla	1,2,3	2007	Fuel: CGH2 provided from AP mobile refueler. Located at HyPoint Industrial Park. Organizations Involved: Air Products, US DOT Research and Innovation Technology Administration, US Air Force Research Laboratory, Defense Logistics Agency.
NV	Las Vegas	Las Vegas Hydrogen Energy Station	1,2,3	2002	Fuels: CGH2, CNG, Hythane. A hydrogen generator produces hydrogen through the reforming and purification of natural gas. The natural gas is provided to the site by a pipeline. Organizations Involved: Air Products and Chemicals Inc., Plug Power Inc., City of Las Vegas.
NV	Las Vegas	Las Vegas Valley Water District, UNLV, DOE	1,2,3	2007	H2 generated by solar-powered electricity. Water district will run two H2 powered trucks and will be adding 8 trucks or shuttles that will use H2 or a blend with natural gas.
NJ	Hopewell	Hopewell Project	2	2006	H2 is generated via electrolysis with electricity from PV panels. H2 is used in a fuel cell powered golf cart.
NM	Taos	Angel's Nest Retreat	2,3	2005	Fuel: CGH2; on-site production with renewable wind and solar power (2 kg of hydrogen per day with 2.5 amps @ 120 v AC; 6 kg storage). Equipment provided by Air Products and Proton Energy Systems.
NY	Albany	Harriman Campus	1,2	2006	Vehicles Served: Fuel cell vehicles, including two Honda FCX's. Organizations Involved: Honda, Air Products, Plug Power, NY State Energy Research and Development Authority (NYSERDA), Homeland Energy.
NY	Ardsley	GM Maintenance and Training Center	2	2008	Station serves GM fuel cell vehicles in the New York City area.
NY	Latham	Home Energy Station, Plug Power	1,2	2004	Reforming via natural gas; able to produce enough hydrogen to fuel one vehicle per day. Vehicles Served: Honda 2005 FCX. Organizations Involved: Plug Power, Honda.
NY	Rochester	Rochester Green City	2	2007	H2 generated by electrolysis with water power. Serves three ICE buses and one fuel cell bus.
NY	White Plains	White Plains Shell Hydrogen	1,2,3	2007	Fuel: CGH2 with 30 kg storage @ 5,000 psi (350 bar) and Hythane. Hydrogen electrolysis from carbon free hydropower source using Distributive Energy Systems FuelGen H2 generator, and Air Products 200H2 fueling technology. Organization Involved: Shell, DES, Air Products, GM, DOE, City of White Plains.

NC	Charlotte	John Deere Southeast Engineering Center Hydrogen Fueling Station	1,3	2004	Provides H2 for FC-powered fork lift. Development includes research and demonstration of pre-commercialized FC powered fork lifts.
ND	Minot	State University North Central Research Extension Center	2,3	2007	Fuel: 80 kg/h CGH2 @ 43 Mpa. Organizations Involved: Hydrogenics, Basic Electric Power Cooperative, Verendrye Electric Cooperative, Velva.
OH	Columbus	Ohio State University	1,2,3	2006	Organizations Involved: Praxair, Ohio State University (OSU), Honda. Ford loaned a Focus for this study; built by OSU Center for Automotive Research.
PA	Allentown	Air Products and Chemicals Headquarters	7	2008	H2 used to fuel two buses, one based at APCI and one at Lehigh Valley Hospital. System capable of storing 140 kg H2.
PA	Penn State University Park	H2VRC H Station	1,2,3	2005	On-site production via natural gas reforming. Vehicles Served: hydrogen vehicles, fuel cell buses. Organizations Involved: Air Products, DOE.
PA	Topton	East Penn Manufacturing Distribution Center	2	2007	On-site production via natural gas reforming using Nuvera Fuel Cells, Inc. PTG-50 hydrogen generation unit. H2 production rate is 2.4 kg/hr. H2 stored in three-bank cascade type system with 6500 psig maximum pressure. H2 used to power fuel cells in forklifts.
SC	A ken	Center for Hydrogen Research	1,3	2008	An Air Prodcuts mobile fueler used to fuel pickup truck ICE. H2 provided at 5000 psig.
VT	Burlington	EVermont Hydrogen Station	1,2,3	2006	Hydrogen used to power a H2 Toyota Prius. Organizations Involved: EVermont, Northern Power Systems, Proton Energy Systems, Air Products, Burlington Department of Public Works, Burlington Electric Department, DoE.
VA	Fort Belvoir	US Army, DoE, General Motors Lead	1,2,3	2004	Fuel: CGH2. Vehicles Served: 40 fuel cell vehicles (Opel Zafira Generation 1 vehicles) for all GM FCV locations. Organizations Involved: GM, Air Products, Linde, U.S. Army, DOE, Shell Hydrogen.

List of Acronyms and Abreviations

APCI	Air Products and Chemicals
ASME	American Society of Mechanical Engineers
AQMD	Air Quality Management District
BAAQMD	Bay Area Air Quality Management District
BP	British Petroleum
CARB	California Air Resources Board
CGH2	Compressed Gaseous Hydrogen
CH2	Compressed Hydrogen (also known as CGH2)
CNG	Compressed Natural Gas
H2	Hydrogen
HCATT	Hawaii Center for Advanced Transportation Technologies
HF 150	APCI brand name for mobile hydrogen fueling station
kg	kilogram
kg/d	kilograms per day
kg/h	kilograms per hour
DOE	(U.S.) Department of Energy
EPA	(U.S.) Environmental Protection Agency
FCV	Fuel Cell Vehicle
LAWA	Los Angeles World Airports
GM	General Motors Corporation
LAX	Los Angeles International Airport
lpm	liter per minute
MPa	Megapascal (35 MPa = 5000 psig =350 bar)
Nm^3/h	Normal cubic meters per hour
PEMFC	Proton Exchange Membrane Fuel Cell
psi	pounds per square inch
psig	pounds per square inch (gauge pressure)
SCAQMD	Southern California Air Quality Management District
SMUD	Sacramento Municipal Utility District
UC	University of California
UNLV	University of Nevada, Las Vegas
VDP	Vehicle Development Program
H2VRC	Hybrid and Hydrogen Vehicle Research Center

Sources of Information about Hydrogen Fueling Stations

1. www.fuelcells.org
2. www.h2stations.org
3. http://www.hydrogenassociation.org/general/fuelingSearch.asp
4. www.cafcp.org
5. http://www.chevron.com/deliveringenergy/hydrogen/
6. http://www.udel.edu/PR/UDaily/2007/apr/bus040907.html
7. http://www.airproducts.com/PressRoom/CompanyNews/Archived/2008/default.htm#October
8. Personal communication with Bruce Russell of Clean Energy.

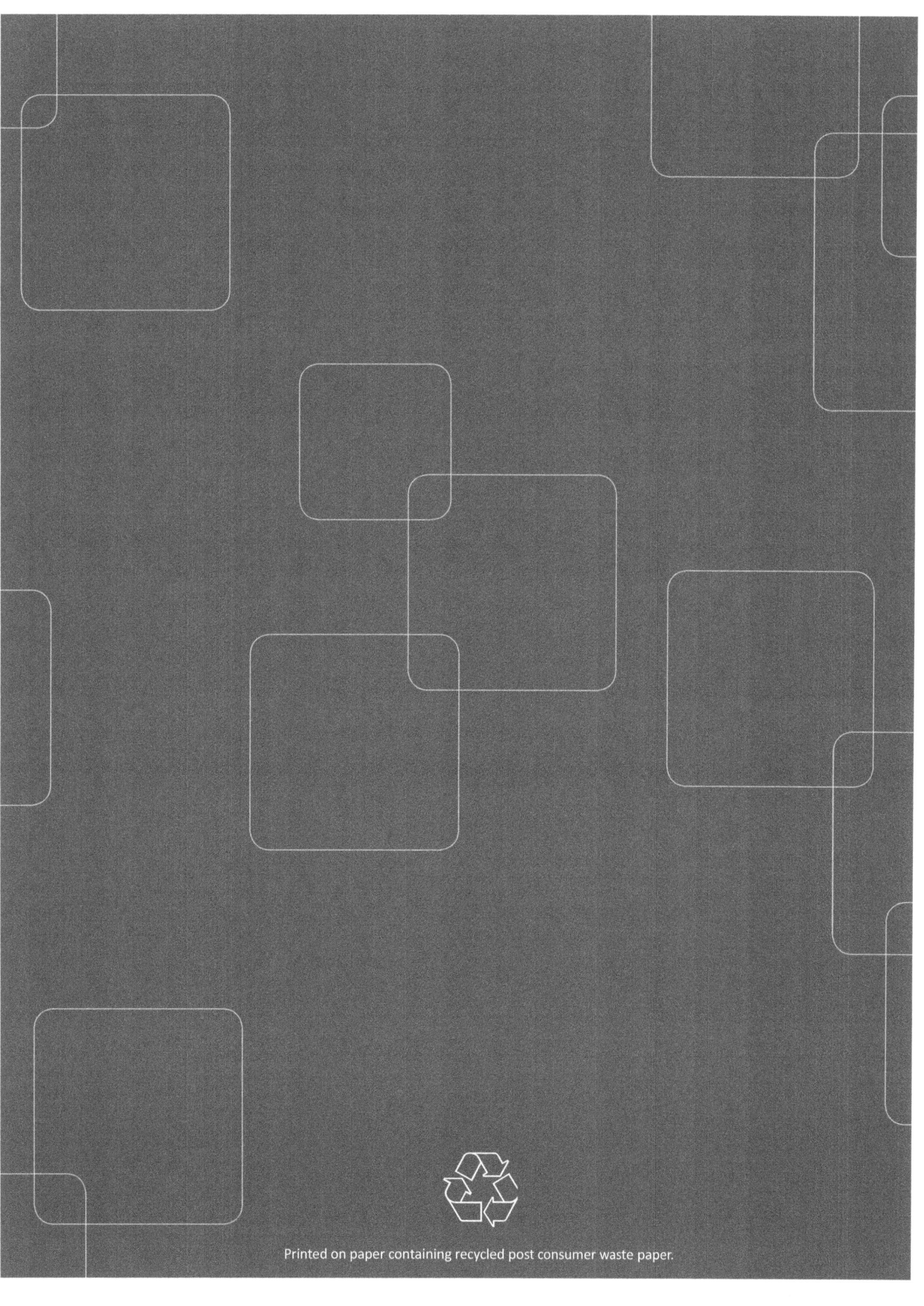
Printed on paper containing recycled post consumer waste paper.

www.ingramcontent.com/pod-product-compliance
Lightning Source LLC
Chambersburg PA
CBHW081856170526
45167CB00007B/3042